Ben Stacy Jerrik (Ed.)

Renard R.35

AF307165

Ben Stacy Jerrik (Ed.)

Renard R.35

Airliner, Monoplane, Sabena, Belgian Congo

Part Press

Contents

Renard_R.35

R.35	
Role	Airliner
National origin	Belgium
Manufacturer	Renard
Designer	Alfred Renard
First flight	1 April 1938
Number built	1

The **Renard R.35** was a prototype pressurised airliner of the 1930s built by the Belgian company Constructions Aéronautiques G. Renard. A three-engined low-winged monoplane with retractable undercarriage, the R.35 was destroyed in a crash on its first flight.

Development and design

In 1935, Alfred Renard, chief designer and co-founder of the Renard company, started design of a pressurised airliner for use by SABENA on its routes to the Belgian Congo. Renard designed a low-winged monoplane of all-metal construction, powered by three engines as required by SABENA, and received an order for a single prototype on 3 April 1936.[1] The R.35 had a circular section fuselage, housing a pressurised cabin which accommodated twenty passengers and a flight crew of three. It was intended to be powered by a range of radial engines with the 950 hp (709 kW) Gnome-Rhône 14K preferred by Renard, but the prototype was fitted with 750 hp (560 kW) Gnome-Rhône 9K engines.[2] [3]

The R.35 was completed early in 1938.[2] On 1 April 1938, it was planned to carry out high-speed taxiing trials at Evere airfield in front of an audience of visiting dignitaries and journalists. After carrying out a single taxi-run, however, the R.35 took off during a second run, and while attempting a circuit to return to the runway, the R.35 dived into the ground and was destroyed, killing the pilot Georges Van Damme. The cause of the crash was unknown.[4] [5] [6] Following this crash, SABENA abandoned its interest in the R.35, and Renard abandoned development.[2]

Variants

R.35B
    Proposed bomber version, capable of carrying 2,800 kg (6,200 lb) of bombs. Unbuilt.[2]

Specifications (performance estimated)

Data from A Belgian Rare Avis [7]

General characteristics

- **Crew:** 3 (2 pilots and radio operator
- **Capacity:** 20 passengers
- **Payload:** 2,000 kg (4,400 lb)
- **Length:** 17.50 m (57 ft 5 in)
- **Wingspan:** 25.50 m (83 ft 8 in)
- **Height:** 5.50 m (18 ft 1 in)
- **Wing area:** 87 m² [3] (936 sq ft)

- **Empty weight:** 6,100 kg (13,400 lb)
- **Max. takeoff weight:** 10,500 kg (23,100 lb)
- **Powerplant:** 3 × Gnome-Rhône 9K 9-cylinder air-cooled radial engine, 560 kW (750 hp) each

Performance

- **Maximum speed:** 435 km/h (235 knots, 270 mph) at 5,000 m (16,400 ft)
- **Cruise speed:** 350 km/h (190 knots, 218 mph)
- **Range:** 1,800 km (974 nmi, 1,120 mi)
- **Service ceiling:** 9,000 m (29,500 ft)

References

Notes

[1] De Wulf 1978, pp. 147-148.
[2] De Wulf 1978, p.148.
[3] Hauet, André. " Renard R.35 Un avion stratosphérique belge en 1938. (technique) (http://aerostories.free.fr/appareils/renard/page2.
html)". *Aérostories*. (In French). Retrieved 26 August 2009.
[4] De Wulf 1978, p.147.
[5] Hauet, André. " Renard R.35 Un avion stratosphérique belge en 1938. (histoire) (http://aerostories.free.fr/appareils/renard/index.html)".
Aérostories. (In French). Retrieved 26 August 2009.
[6] " ASN Aircraft accident 01-APR-1938 Renard R.35 OO-ARM (http://aviation-safety.net/wikibase/wiki.php?id=17301)". *Aviation Safety
Network*. 15 April 2009. Retrieved 26 August 2009.
[7] De Wulf 1978, p.149.

Bibliography

- De Wulf, Herman. "A Belgian Rare Avis". *Air International*, September 1978, Vol 15 No. 3. Bromley, UK: Fine
  Scroll. pp. 147-149.

External links

- photo (http://www.baha.be/Webpages/Navigator/Photos/CivilPics/civil_pics_ooaaa_ooczz/RenardR35.
  htm)
- RENARD R-35 (http://www.fnar.be/html/R35/R-35.htm)

Airliner

Airliner

The Boeing 747 was the first wide-body airliner. Pictured is the 747-8I model which first flew in 2011.

An **airliner** is a large fixed-wing aircraft for transporting passengers and cargo. Such aircraft are operated by airlines. Although the definition of an airliner can vary from country to country, an airliner is typically defined as an aircraft intended for carrying multiple passengers in commercial service.

The Airbus A320 narrow-body is a popular short-medium distance aircraft.

History

A United Airlines DC-6 at Stapleton Airport, Denver, in September 1966

When the Wright brothers made the world's first sustained heavier-than-air flight, they laid the foundation for what would become a major transport industry. Their flight in 1903 was just 11 years before what is often defined as the world's first airliner.[1] These airliners would change the world socially, economically, and politically in a way that had never been done before.[2]

If an airliner is defined as a plane intended for carrying multiple passengers in commercial service, the Russian Sikorsky Ilya Muromets was the first official passenger aircraft. The Ilya Muromets was a luxurious aircraft with an isolated passenger saloon, wicker chairs, bedroom, lounge and a bathroom. The aircraft also had heating and electrical lighting. The Ilya Muromets first flew on December 10, 1913. On February 25, 1914, it took off for its first demonstration flight with 16 passengers aboard.

From June 21 – June 23, it made a round-trip from Saint Petersburg to Kiev in 14 hours and 38 minutes with one intermediate landing. If it had not been for World War I, the Ilya Muromets would have probably started passenger flights that same year.

The second airliner was the Farman F.60 Goliath from 1919, which could seat up to 14 passengers, approximately 60 were built. The world's first all-metal aircraft was the Junkers F.13, also from 1919 with 322 built.

The Ford Trimotor was an important early airliner. With two engines mounted on the wings and one in the nose and a slabsided body, it carried eight passengers and was produced from 1925 to 1933. It was used by the predecessor to TWA as well as other airlines long after production ceased. In 1932 the 14-passenger Douglas DC-2 flew and in 1935 the more powerful, faster, 21–32 passenger Douglas DC-3. DC-3s were produced in quantity for WWII and sold as surplus afterward.The Douglas DC-3 was a particularly important airplane because it was the first airliner to be profitable without a government subsidy.[3]

The first jet airliners came in the immediate post war era. Turbojet engines were trialled on piston engine airframes such as the Avro Lancastrian and the Vickers VC.1 Viking the latter becoming the first jet engined passenger aircraft in April 1948. The first purpose built jet airliners were the de Havilland Comet (UK) and the Avro Jetliner (Canada). The former entered production and service while the latter did not. The Comet was unfortunate in that metal fatigue caused by the square shape of the windows in early versions could cause crashes.

Jets did not immediately replace piston engines and many designs used the turboprop rather than the turbojet or the later turbofan engines.

Postwar airliner history in the United States

The United States gained a huge advantage in design and production in the airline industry in the years leading up to the war, but many of the developments would be put off until after the war as the manufacturing efforts were placed on the war effort. The advancements that the United States would make in this industry were in large due to the cooperation of the airlines discussing what they desired with the airliner manufacturers. Soon after the war though Douglas made a large advancement with the DC-4, although this could not cross the Atlantic at every point, it was able to make a nonstop flight from New York to the United Kingdom. Due to the war going on, the first batch of these planes went to the US Army and Air Forces, and was named the C-54 Skymaster. Some of these that were used in the war would later be converted for the airline industry, along with the passenger and cargo versions that were placed on the market once the war ended. Douglas would later develop a version of this plane that was pressurized and 5 feet longer; this redesigned plane would become the DC-6. These DC-6s would be grounded for 6 months to rectify a few safety issues that were causing in-flight fires.

Soon after the DC-4, Lockheed developed the distinctive tri-tail Constellation. An aviation breakthrough, it was the first pressurized airliner, allowing it to fly higher, and therefore further and faster than ever before. Its fuselage was some 127 inches wider than the DC-4's. Drafted by the military in WWII, it experienced a similar late entry into the civilian airline industry. Safety concerns grounded it for six months soon after it entered service while problems were investigated and repaired.

In 1947 the Boeing 377 Stratocruiser entered the industry with a completely different design than Douglas and Lockheed aircraft. Based on the C-97 military transport, it had a double deck and pressurized fuselage. Luxury and a 100 passenger capacity distinguished it from its rivals. While 900 C-97s were supplied to the military only 55 were produced for civil aviation.

The American companies had done a great job of advancing the status of transcontinental travel, but there was also the aging fleet of DC-3s that had to be addressed. Convair decided that they were going to address this market, and would begin producing the Convair 240 which was a 40 person fully pressurized plane. There were 566 of these planes that would fly, including 2 that were equipped with jet-assisted take off units. Convair would later develop the Convair 340, which was slightly larger and could accommodate between 44 and 52 passengers, and 311 of this model plane were produced. Finally Convair would create a Convair 440, which had small modifications including much better soundproofing than the previous models. Convair would experience a little bit of competition from the Martin 2-0-2 and Martin 4-0-4, but in general Convair was able to control this market, as the 2-0-2 had safety concerns and was unpressurized, and the 4-0-4 only sold around 100 units.[4]

The United States was dominant in this industry for several reasons including a large domestic market for these planes. The market would also work in the United States favor as the American companies began to build pressurized airliners. During the postwar years engines became much larger and more powerful, and safety features such as deicing, navigation, and weather were added to the planes. Lastly, the planes produced in the United States were more comfortable and had superior flight decks than those produced in Europe.[4]

Postwar airliner history in Great Britain

Great Britain was in a very different position after the war than the United States. Unlike the United States, Great Britain had a small domestic market and almost all of the airplane construction that had take place domestically was for war. In December 1942 the British government had a committee put in place to set classifications for airplanes ranging from Non-Stop North Atlantic airplanes to Small Piston-engine airplanes for light traffic. These classifications were set up to encourage development of airliners of all types. In order to recover from the difficulties that this caused many of the postwar airliners were bombers that were converted to allow for commercial air travel, but this was not very economical as the planes could hold very few people. The first postwar program that was attempted in Great Britain was the Tudor airliner, however this was regarded as a failure due to safety concerns and few sales. The first successful British Airliner was a Vickers Armstrongs Viking. These were unpressurized and could hold between 21 and 27 people depending on the model. On April 6, 1948 one of these Viking airliners became the first jet-airliner to fly. The previous engines were now replaced with Rolls-Royce Nene Turbojets. In 1946 the Bristol 170 was the first transport aircraft to receive a Certificate of Airworthiness from the British government. This plane remained in production for 12 years with 214 of the aircraft built. The British would also produce several smaller aircraft such as the Airspeed AS.57 Ambassador in 1947 and the Miles M.57 Aerovan in 1945, and the DH Dove and its four-engined cousin the Heron. Most of the British planes produced in the postwar era were smaller planes that carried less than 30 passengers, and did not sell as well as some of the planes produced in the United States, but these planes were enough to help the British airline industry from the lack of commercial production during the war. [4]

Postwar airliner history in France

In the postwar years France developed a few significant airliners, some of these being planes that could land on water, part of the reason that the French companies were so focused on these flying boats is that in 1936 the French Air Ministry requested transatlantic flying boats that could hold at least 40 passengers. Only one model from this request would ever be put into service. The first set of these was 3 Latecoere 631's that Air France purchased and put into service in July 1947. However, two of these planes crashed, and the third plane was soon removed because of these safety concerns. There would later be a SNCASE SE.161 Languedoc build, which was a much more successful plane, and over 100 of these were built, with 40 of them being placed into service through Air-France.[4] The French also developed the Breguet 763 Deux Ponts, which first flew in February 1949 . This was a double-decker transport airliner that would end up being used for both people and cargo. This four-engine airliner would end up being used to hold massive amounts of cargo or 97 passengers. After a long silence, France then created the Caravelle, the world's first short-to-midrange jet airliner. Subsequent French efforts were part of the Airbus pan-European initiative.

Postwar airliner history in the USSR

Soon after the war most of the Soviet fleet of airliners consisted of DC-3s or the Lisunov Li-2. These planes were in desperate need of replacement, and in 1946 the Ilyushin Il-12 made its first flight. The Il-12 was very similar in design to American Convair 240, except was unpressurized. In 1953 the Ilyushin Il-14 would make its first flight, and this version was equipped with much more powerful engines.The main contribution that the Soviets made in regards to Airliners was the Antonov An-2. This plane is a biplane unlike most of the other airliners and sold more units than any other transport plane.[4]

Types

Wide-body airliners

The largest airliners are *wide-body* jets. These aircraft are frequently called *twin-aisle aircraft* because they generally have two separate aisles running from the front to the back of the passenger cabin. Aircraft in this category are the Boeing 747, Boeing 767, Boeing 777, Boeing 787, Airbus A300/A310, Airbus A330, Airbus A340, Airbus A380, Lockheed L-1011 TriStar, McDonnell Douglas DC-10, McDonnell Douglas MD-11, Ilyushin Il-86 and Ilyushin Il-96. These aircraft are usually used for long-haul flights between airline hubs and major cities with many passengers. Future wide-body models include the Airbus A350.

The Airbus A330 is a wide-body airliner

Narrow-body airliners

The Boeing 757-300 is the longest narrow-body airliner ever built.

A smaller, more common class of airliners is the *narrow-body* or *single aisle* aircraft. These smaller airliners are generally used for medium-distance flights with fewer passengers than their wide-body counterparts.

Examples include the Boeing 717, 737, 757, McDonnell Douglas DC-9 and MD-80/MD-90 series, Airbus A320 family, Tupolev Tu-204, Tu-214, Embraer E-Jets 190&195 and Tu-334. Older airliners like the Boeing 707, 727, Douglas DC-8, Fokker F70/F100, VC10, Tupolev, and Yakovlev jets also fit into this category.

Small airliners

The Boeing 737 is the most mass-produced narrow-body of all time.

Short haul airliners used by airlines and regional airlines

A JetBlue Airways Embraer 190 short haul airliner.

Regional airliners - Small (Regional) short haul airliners typically seat fewer than 100 passengers and may be powered by turbofans or turboprops.

A PLUNA Bombardier CRJ900 short haul (regional) airliner taxiing.

These airliners, although smaller than aircraft operated by most major carriers, legacy carriers, or flag carriers, frequently serve customers who expect service similar to that offered by the larger airlines with their longer-ranged, larger airliners.

Therefore, these short-haul airliners are usually equipped with lavatories, stand up cabins, pressurization, overhead storage bins, reclining seats, and have a flight attendant to look after the in-flight needs of the passengers during point-to-point or routes.

Direktflyg Jetstream 32 at Kristiansund Airport, Kvernberget

Because these aircraft are frequently operated by smaller airlines that are contracted to provide ("feed") passengers from smaller cities to hub airports (and reverse) for a "major" or "flag" carrier, regional airliners may be painted in the liveries of the major airline for whom they provide this "feeder" service. (See below)

Feederliner aircraft used by regional airlines

Regional airliners - (Regional) Feederliners typically seat fewer than 100 passengers and may be powered by turbofans or turboprops. These airliners, are the non mainline counterparts to the larger aircraft operated by the; major carriers, legacy carriers, and flag carriers and are used to feed traffic into the large airline hubs or focus cities. These particular routes may need the size of a smaller aircraft to meet the frequency needs and service levels, customers expect in the marketed product that is offered by larger airlines and their modern narrow and

The Bombardier CRJ200

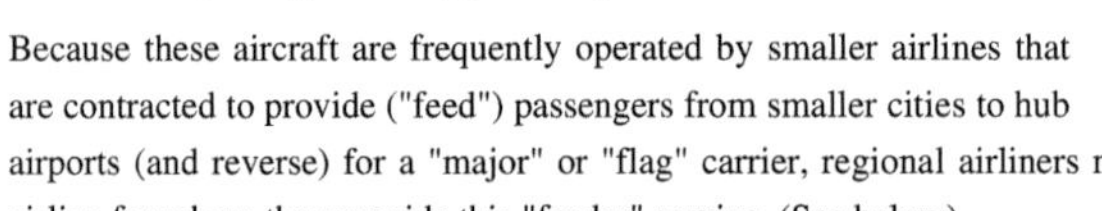

widebody aircraft. Therefore, most regional airliners are equipped with lavatories and have a flight attendant to look after the in-flight needs of the passengers, along with the features of a short haul regional airliner.

Typical aircraft in this category include the Bombardier CRJ and Embraer ERJ regional jets along with the "Q" (DASH-8) series, ATR 42/72 and Saab 340/2000 turboprop airliners. Airlines and their partners sometimes use these for flights between small hubs, or for bringing passengers to hub cities where they may board larger aircraft. Typically, these regional feederliners, are painted in the aircraft liveries and color schemes of the much larger airline partners so the regional airlines may offer and market a seamless transition between the larger airline to smaller airline.

A Compass Airlines (North America) Embraer ERJ-170-200LR in the feederliner colors of Northwest Airlink

Commuterliner aircraft used by regional airlines and air taxi charter operators

The Beechcraft 1900 short range commuter aircraft

The lightest (light aircraft, list of light transport aircraft) of short haul regional feeder airliner type aircraft that carry 19 or fewer passenger seats are called *commuter aircraft, commuterliners, feederliners,* and *air taxis*, depending on their size, engines, how they are marketed, region of the world, and seating configurations. The Beechcraft 1900, for example, has only 19 seats. Depending on local and national regulations, a commuter aircraft may not qualify as an airliner and may not be subject to the regulations applied to larger aircraft. Members of this class of aircraft normally lack such amenities as lavatories and galleys and typically do not carry a flight attendant as an aircrew member.

Other aircraft that may fall into this category are the Fairchild Metro, Jetstream 31, and Embraer EMB 110 Bandeirante. The Cessna Caravan and Pilatus PC-12, are single-engine turboprops, sometimes used as a small airliner, although many countries stipulate a minimum requirement of two engines for aircraft to be used as airliners.

Twin piston-engined aircraft made by Cessna, Piper, Britten-Norman, and Beechcraft are also in use as short haul, short range commuter type aircraft.

Engines

Until the beginning of the Jet Age, piston engines were common on propliners like the Douglas DC-3. Nearly all modern airliners are now powered by turbine engines, either turbofans or turboprops. Gas turbine engines operate efficiently at much higher altitudes, are more reliable than piston engines, and produce less vibration and noise. Prior to the Jet Age, it was common for the same or very similar engines to be used in civilian airliners as in military aircraft. In recent years, divergence has occurred so that it is now unusual for the same engine to be used on a military type as a civilian type. Usually military aircraft which share engine technology with airliners are transports or tanker types.

Airliner variants

Some variants of airliners have been developed for carrying freight or for luxury corporate use. Many airliners have also been modified for government use as VIP transports and for military functions such as airborne tankers (for example, the Vickers VC10, Lockheed L1011, Boeing 707), air ambulance (USAF/USN McDonnell Douglas DC-9), reconnaissance (Embraer ERJ 145, Saab 340, Boeing 737), as well as for troop-carrying roles.

Configuration

Modern airliners are usually low-wing designs with two engines mounted in underwing pods. The Boeing 747 and Airbus A380 are the only airliners in production which are too heavy (more than 400 tons maximum takeoff weight) for just two engines. Smaller airliners sometimes have their engines mounted on either side of the rear fuselage. There are numerous advantages and disadvantages to this arrangement.[5] Perhaps the most important advantage to mounting the engines under the wings is that the total aircraft weight is more evenly distributed across the wingspan, which imposes less bending moment on the wings and allows for a lighter wing structure. This factor becomes more important as aircraft weight increases, and there are no in-production airliners with both a maximum takeoff weight of more than 50 tons and engines mounted on the fuselage. The Antonov An-148 is the only in-production airliner with high-mounted wings (usually seen in military transport aircraft), which reduces the risk of damage from unpaved runways.

Except for a few experimental or military designs, all aircraft built to date have had all of their weight lifted off the ground by airflow across the wings. In terms of aerodynamics, the fuselage has been a mere burden. NASA and Boeing are currently developing a blended wing body design in which the entire airframe, from wingtip to wingtip, contributes lift. This promises a significant gain in fuel efficiency.[6]

Manufacturers

These include:

* Asia
    * China
        * COMAC (includes Shanghai Aircraft Manufacturing Factory)
        * Shenyang Aircraft Corporation
        * Xi'an Aircraft Industrial Corporation
* Europe
    * Airbus S.A.S. (formerly a multinational conglomeration of the largest European aerospace companies of France, Germany, Spain and the UK)
    * Czech Republic
        * Let Kunovice
    * France/Italy
        * ATR
    * Netherlands
        * Fokker (now defunct)
    * Russian companies (formerly Soviet-controlled)
        * Ilyushin
        * Sukhoi
        * Tupolev

Assembly of a Boeing 767 airliner nose section

- Yakovlev
- Sweden
  - Saab AB (no longer manufactures civilian aircraft)
- Ukraine (formerly Soviet-controlled)
  - Antonov
- United Kingdom
  - BAE Systems (formerly British Aerospace, no longer manufactures civilian aircraft)
  - Britten-Norman
- North America
  - Canada
    - Bombardier (includes the former De Havilland Canada and Canadair)
  - United States
    - Boeing (includes the former McDonnell Douglas company which itself included the Douglas Aircraft Company)
    - Lockheed Corporation (now part of Lockheed Martin, and no longer involved in civil aviation)
- South America
  - Brazil
    - Embraer

The international market for middle-sized and large-sized airliners is now divided between Airbus and Boeing, although Russian/former Soviet manufacturers still sell significant numbers of airliners to their traditional markets. Smaller-sized aircraft manufacturers include, in addition to these two, ATR, Embraer and Bombardier.

Notable airliners

- Boeing 247 – the first design to incorporate modern features such as all-metal construction and retractable landing gear
- Douglas DC-3 – still in service more than 70 years after its debut, it is generally regarded as one of the most significant transport aircraft ever made
- Boeing 307 - One of the first airliners to be pressurized and have the ability to fly into the stratosphere
- Douglas DC-6 – originally developed as a military transport, it was reworked for passenger service after World War II, a role it continues to perform today

Notable airliners – an Airbus A380 "superjumbo" of Singapore Airlines takes off

- Boeing 377- Developed soon after World War 2, from the C-97 Stratofreighter, this was a luxurious double-decker airliner with a pressurized cabin.
- Vickers Viscount – the first turboprop airliner to enter service
- Lockheed Constellation – a distinctive triple-tailed piston-engined airliner of the 1950s, it was one of the last large propeller-driven airliners
- De Havilland Comet – the world's first jetliner to reach mass production, its reputation was marred by a series of crashes due to structural failure
- Antonov An-2- Best selling transport airliner up to the point it was built.
- Tupolev Tu-104 – the first turbojet airliner to provide sustained service, and the sole jetliner operating in the world between 1956 and 1958
- Boeing 707 – the first United States-built jetliner to enter production

- Douglas DC-8 – launched after the Boeing 707, it nevertheless established Douglas in the airliner market, and continues to serve as a cargo aircraft to this day
- Tupolev Tu-154 - standard medium-range airliner for Russia (and others), carried half of all Soviet traffic since 1972 with 1015 built and the fastest airliner in service
- Ilyushin Il-62 - standard long-range airliner for Russia (and others) for three decades, first flight 1963 and still in service
- Douglas DC-9 – production of it and successive variants nearly reached 2,500
- Boeing 737 – the best-selling jet airliner in the history of aviation[7]
- Tupolev Tu-144 – the first supersonic transport aircraft constructed in Soviet Union
- Concorde – an Anglo-French supersonic transport, it remains the only supersonic aircraft to sustain a regular passenger service
- Boeing 747 "jumbo jet" – an iconic aircraft, it was the world's largest airliner between 1969 and 2005
- McDonnell Douglas DC-10 – a trijet competitor to the widebody 747
- Airbus A300 – the world's first twinjet widebody
- Airbus A320 – pioneered the use of fly-by-wire technology
- Boeing 777 – the first airliner designed entirely by computer, without physical mockups
- Airbus A380 "superjumbo" – the world's largest airliner from 2005 onwards
- Boeing 787 Dreamliner - the world's first jet airliner to make use of composite materials for most of its construction

Airliner recycling

As airliners are very expensive, most are leased out for times typically from 20 to 40 years. Very few go back into service after a long lease is up because evolving aerospace technology leaves older airliners unable to compete against newer machines that can be operated at a lower cost. Many end-of-service airliners end up in the Mojave Desert, at the Mojave Air and Space Port (also known as "The Boneyard"). From this, the term "Mojave" has come to refer to the temporary storage of aircraft, e.g. during decreased demand for air travel and between short-term leases. Another airliner retirement location is Marana, Arizona.

While almost every airliner will be reduced to scrap (the exceptions end up as museum pieces or flown by collector groups) they may pass through many owners before they are retired. A well-maintained airliner can operate safely for decades, depending on how often it is flown, its operating environment, and whether damage and wear and tear is properly repaired.

What may end an airliner's working life is a lack of spare parts, as the original manufacturer and third manufacturers may no longer provide or support them. Corrosion and metal fatigue are other issues that become more expensive to deal with as time goes on. Eventually, these factors and advances in aircraft technology lead to older airliners becoming too expensive or inefficient to operate.

To protect the environment, the Airbus company has set up a centre in France to decommission and recycle older aircraft. More than 200 airliners will finish active life each year, and will be dismantled and recycled under the newly established PAMELA Project.

Cabin configurations and features

An airliner will usually have several classes of seating: first class, business class, and/or economy class (which may be referred to as coach class or tourist class, and sometimes has a separate "premium" economy section with more legroom and amenities). The seats in more expensive classes are wider, more comfortable, and have more amenities such as "lie flat" seats for more comfortable sleeping on long flights. Generally, the more expensive the class, the better the beverage and meal service.

Interior of a Qatar Airways Airbus. Video systems (the vertical white panels) are visible above the very centre seats of the aircraft

Domestic flights generally have a two-class configuration, usually first or business class and coach class, although many airlines instead offer all-economy seating. International flights generally have either a two-class configuration or a three-class configuration, depending on the airline, route and aircraft type. Many airliners offer movies or audio/video on demand (this is standard in first and business class on many international flights and may be available on economy). Cabins of any class are provided with lavatory facilities.

Seats

The types of seats that are provided and how much legroom is given to each passenger are decisions made by the individual airlines, not the aircraft manufacturers. Seats are mounted in "tracks" on the floor of the cabin and can be moved back and forth by the maintenance staff or removed altogether. Naturally the airline tries to maximize the number of seats available in every aircraft to carry the largest possible (and therefore most profitable) number of passengers.

Boarding an Airbus A380 at the Farnborough Airshow, 2006

Passengers seated in an **exit row** (the row of seats adjacent to an emergency exit) usually have substantially more legroom than those seated in the remainder of the cabin, while the seats directly in front of the exit row may have less legroom and may not even recline (for evacuation safety reasons). However, passengers seated in an exit row may be required to assist cabin crew during an emergency evacuation of the aircraft opening the emergency exit and assisting fellow passengers to the exit. As a precaution, many airlines prohibit young people under the age of 15 from being seated in the exit row [8].

The seats are designed to withstand strong forces so as not to break or come loose from their floor tracks during turbulence or accidents. The backs of seats are often equipped with a fold-down tray for eating, writing, or as a place to set up a portable computer, or a music or video player. Seats without another row of seats in front of them have a tray that is either folded into the armrest or that clips into brackets on the underside of the armrests. However, seats in premium cabins generally have trays in the armrests or clip-on trays, regardless of whether there is another row of seats in front of them. Seatbacks now often feature small color LCD screens for videos, television and video games. Controls for this display as well as an outlet to plug in audio headsets are normally found in the armrest of each seat.

Overhead bins

The overhead bins are used for stowing carry-on baggage and other items. While the airliner manufacturer will normally supply a standard product, airlines may choose to have bins of differing size, shape, or color installed. Over time, these bins evolved out of what were originally overhead shelves used for little more than coat and briefcase storage. As concerns about falling debris during turbulence or in accidents increased, enclosed bins became the norm. Bins have increased in size in order to accommodate the larger carry-on baggage passengers may bring onto the aircraft. New bin designs may include a handrail, useful when moving through the cabin.

Passenger service units

Above the passenger seats are Passenger Service Units (PSU). These typically contain reading lights, air vents, and a flight attendant call light. On most narrowbody aircraft (and some Airbus A300s and A310s), the flight attendant call button and the buttons to control the reading lights are located directly on the PSU, while on most widebody aircraft, the flight attendant call button and the reading light control buttons are usually part of the in-flight entertainment system. The units frequently have small "Fasten Seat Belt" and "No Smoking" illuminated signage and may also contain a speaker for the cabin public address system.

The PSU will also normally contain the drop-down oxygen masks which are activated if there is a sudden drop in cabin pressure. These are supplied with oxygen by means of a chemical oxygen generator. By using a chemical reaction rather than a connection to an oxygen tank, these devices supply breathing oxygen for long enough for the airliner to descend to thicker, more breathable air. Oxygen generators do generate considerable heat in the process. Because of this, the oxygen generators are thermally shielded and are only allowed in commercial airliners when properly installed – they are not permitted to be loaded as freight on passenger-carrying flights. ValuJet Flight 592 crashed on May 11, 1996 as a result of improperly loaded chemical oxygen generators.

Cabin pressurization

Airliners developed since the 1940s have had pressurized cabins (or more accurately, pressurized hulls including baggage holds) to enable them to carry passengers safely at high altitudes where low oxygen levels and air pressure would otherwise cause sickness or death. High altitude flight enabled airliners to fly above most weather systems that cause turbulent or dangerous flying conditions, and also to fly faster and further as there is less drag due to the lower air density. Pressurisation is applied using compressed air, in most cases bled from the engines, and is managed by a environmental control system which draws in clean air, and vents stale air out through a valve.

Pressurization presents design and construction challenges to maintain the structural integrity and sealing of the cabin and hull and to prevent rapid decompression. Some of the consequences include small round windows, doors that open inwards and are larger than the door hole, and an emergency oxygen system.

To maintain a pressure in the cabin equivalent to an altitude close to sea level would, at a cruising altitude around 10,000 m (33,000 feet), create a pressure difference between inside the aircraft and outside the aircraft that would require greater hull strength and weight. Most people do not suffer ill effects up to an altitude of 1800–2500 m (6000–8000 feet), and maintaining cabin pressure at this equivalent altitude significantly reduces the pressure difference and therefore the required hull strength and weight. A side effect is that passengers experience some discomfort as the cabin pressure changes during ascent and descent to the majority of airports, which are at low altitudes.

Cabin climate control

The air bled from the engines is hot and requires cooling by air conditioning units. It is also extremely dry at cruising altitude, and this causes sore eyes, dry skin and mucosa on long flights. Although humidification technology could raise its relative humidity to comfortable middle levels, this is not done since humidity promotes corrosion to the inside of the hull and risks condensation which could short electrical systems, so for safety reasons it is deliberately kept to a low value, around 10%.

Baggage holds

Airliners must have space on board to store baggage that will not safely fit in the passenger cabin.

Designed to hold baggage as well as freight, these compartments are called "cargo bins", "holds", or occasionally "pits". Occasionally baggage holds may be referred to as **cargo decks** on the largest of aircraft. These compartments can be accessed through doors on the outside of the aircraft. Despite what is seen in many movies, access doors between passenger cabins and baggage holds are rare in modern airliners.

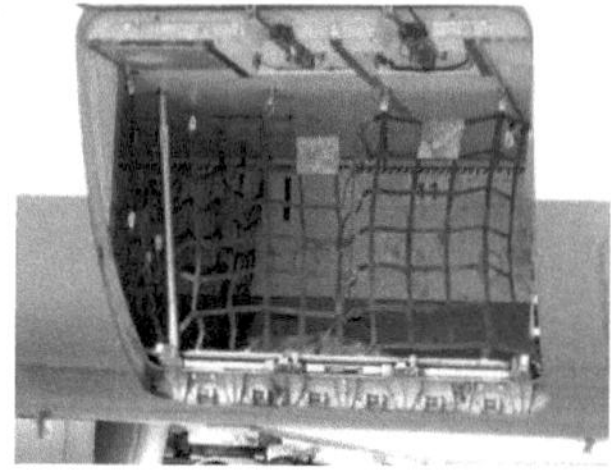

An Airbus A320 baggage hold

Depending on the aircraft, baggage holds are normally inside the hull and are therefore pressurized just like the passenger cabin although they may not be heated. While lighting is normally installed for use by the loading crew, typically the compartment is unlit when the door is closed.

Baggage holds on modern airliners are equipped with fire detection equipment and larger aircraft have automated or remotely activated fire-fighting devices installed.

Narrow-body airliners

Most "narrow-body" airliners with more than 100 seats have space below the cabin floor, while smaller aircraft often have a special compartment separate from the passenger area but on the same level.

Loading luggage onto a Boeing 747 at Boston Logan Airport, during a closure due to heavy snow

Baggage is normally stacked within the bin by hand, sorted by destination category. Netting that fits across the width of the bin is secured to limit movement of the bags. Airliners often carry items of freight and mail. These may be loaded separately from the baggage or mixed in if they are bound for the same destination. For securing bulky items "hold down" rings are provided to tie items into place.

Wide-body airliners

"Wide-body" airliners frequently have a compartment like the ones described above, typically called a "bulk bin". It is normally used for late arriving luggage or bags which may have been checked at the gate.

Boeing 747 front lower compartment. Note the rollers for ULDs on the floor and the partition labeled "Caution: Do Not Hit -- Potable Water Tank Inside".

However, most baggage and loose freight items are loaded into containers called Unit Load Devices (ULDs), often referred to as "cans". ULDs come in a variety of sizes and shapes, but the most common model is the LD3. This particular container has approximately the same height as the cargo compartment and fits across half of its width.

ULDs are loaded with baggage and are transported to the aircraft on dolly carts and loaded into the baggage hold by a loader designed for the task. By means of belts and rollers an operator can maneuver the ULD from the dolly cart, up to the aircraft baggage hold door, and into the aircraft. Inside the hold, the floor is also equipped with drive wheels and rollers that an operator inside can use to move the ULD properly into place. Locks in the floor are used to hold the ULD in place during flight.

For consolidated freight loads, like a pallet of boxes or an item too oddly shaped to fit into a container, flat metal pallets that resemble large baking sheets that are compatible with the loading equipment are used.

See also

Lists

- Regional jets
- List of civil aircraft
- List of regional airliners
- List of airliners by Maximum Takeoff Weight

Topics

- Airport
- Air travel
- Air safety
- Flight planning
- Aircraft spotting
- Aviation and the environment

References

Notes

[1] Tise, Larry E. Conquering the Sky. New York: Palgrave MacMillan, 2009. Print.

[2] Bernardo, James V. Aviation and Space: In the Modern World. New York: E.P Dutton & Co., Inc., 1968. Print.

[3] Mellberg, William F. 2003. "TRANSPORTATION REVOLUTION." Mechanical Engineering 125, 22-25. Academic Search Premier, EBSCOhost (accessed October 19, 2010).

[4] Jarrett, Phillip. eds. Modern Air Transport: Worldwide Air Transport from 1945 to the Present. London: Putnam, 2000.

[5] Kroo, Ilan (19 January 2006). "Engine Placement" (http://adg.stanford.edu/aa241/propulsion/engineplacement.html). *AA241 Introduction to Aircraft Design: Synthesis and Analysis*. Stanford University. . Retrieved 12 February 2012.

[6] Warwick, Graham (13 January 2009). "NASA Pushes Blended Wing/Body" (http://www.aviationweek.com/aw/generic/story. jsp?id=news/Body011309.xml&headline=NASA Pushes Blended Wing/Body &channel=space). *Aviation Week*. . Retrieved 12 February 2012.

[7] Kingsley-Jones, Max. "6,000 and counting for Boeing's popular little twinjet." (http://www.flightglobal.com/articles/2009/04/22/ 325472/pictures-6000-and-counting-for-boeings-popular-little-twinjet.html) *Flight International*, Reed Business Information, April 22, 2009. Retrieved: April 22, 2009.

[8] http://www.casa.gov.au/airsafe/trip/seating.htm

http://www.grc.nasa.gov/WWW/K-12/airplane/turbine.html

"Aviation Industry." International Encyclopedia of the Social Sciences. 2008. Retrieved March 25, 2011 from Encyclopedia.com: http://www.encyclopedia.com/doc/1G2-3045300147.html

Brayton thermodynamic cycle http://www.grc.nasa.gov/WWW/K-12/airplane/brayton.html

Bibliography

- Newhouse, John. *The Sporty Game: The High-Risk Competitive Business of Making and Selling Commercial Airliners.* New York: Alfred A. Knopf, 1982. ISBN 978-0-394-51447-5.

External links and references

- Boeing (http://www.boeing.com)
- Airbus (EADS) (http://www.airbus.com)
- International Lease Finance Corporation (http://www.ilfc.com)
- Embraer (http://www.embraer.com)
- Bombardier (http://www.bombardier.com)
- ATR (http://www.atraircraft.com)
- Canadian Air Transport Security Authority (CATSA) website (http://www.catsa-acsta.gc.ca/)
- Airliners.net (http://www.airliners.net)
- Airplanes.se (http://www.airplanes.se)

Monoplane

For Félix du Temple's invention, see Monoplane (1874)

Monoplane

The low-wing of a Curtiss P-40

A **monoplane** is a fixed-wing aircraft with one main set of wing surfaces, in contrast to a biplane or triplane. Since the late 1930s it has been the most common form for a fixed wing aircraft.

Types of monoplane

The main distinction between types of monoplane is where the wings attach to the fuselage:

The mid-wing of a de Havilland Vampire T11.

- **low-wing**, the wing lower surface is level with (or below) the bottom of the fuselage
- **mid-wing**, the wing is mounted mid-way up the fuselage
- **shoulder wing**, the wing is mounted above the fuselage middle
- **high-wing**, the wing upper surface is level with or above the top of the fuselage
- **parasol-wing**, the wing is located above the fuselage and is not directly connected to it, structural support being typically provided by a system of struts, and, especially in the case of older aircraft, wire bracing.

The shoulder-wing of an ARV Super2.

History

Most of the first attempts of heavier-than-air flying machines were monoplanes. Notably, the *Monoplane* built in 1874 by Felix du Temple de la Croix, a large aircraft made of aluminium with a wingspan of 13 m (**unknown operator: u'strong'** ft) and a weight of only 80 kg (**unknown operator: u'strong'** lb) (without the pilot). Several trials were made with the plane in Brest, France, and it is generally recognized that it achieved lift off under its own power after a ski-jump run, glided for a short time and returned safely to the ground, possibly making it the first successful powered flight in history, depending on the definition — since the flight was only a few feet high, and was not truly under control.

Other early attempts of flight by a monoplane were carried out in 1884 by Alexander Mozhaysky.[1]

The first successful aircraft were biplanes, but many pioneering aircraft were monoplanes, for instance Blériot XI that flew across the English Channel in 1909. Throughout 1909-1910 Hubert Latham set multiple altitude records in his Antoinette IV monoplane, initially achieving 155 m (**unknown operator: u'strong'** ft) then raising it to 1384 m (**unknown operator: u'strong'** ft).[2] The Fokker Eindecker of 1915 was a successful fighter aircraft. The Junkers J 1 was an early German "technology demonstrator" monoplane, and the world's very first practical all-metal aircraft of any type to fly, with the J 1's first flight occurring in December 1915.

According to the Encyclopædia Britannica, the first successful monoplane flight – a duration of 12 m (40 feet) at Montesson, France – was on March 18, 1906 in a craft fashioned by Traian Vuia, a Romanian inventor.[3] [4]

Nonetheless, relatively few monoplane types were built between 1914, and the late 1920s, compared with the number of biplanes. The reasons for this were primarily structural. In the days when wings (whether biplane or monoplane) were thin, lightly built structures, braced by struts, steel wire or cables - the biplane wing formed a strong and fairly rigid lattice truss structure, in which the two wing surfaces were braced against each other. Early monoplane wings, on the other hand, tended to be liable to twist under aerodynamic loads, rendering proper lateral control very difficult. They were also much more liable to breakage in flight.

Once all-metal construction and the cantilever wing, both having been pioneered by Hugo Junkers in 1915, became common after World War I's end, however, the day of the biplane very quickly passed, and the monoplane became the usual configuration for a fixed-wing aircraft.

Most military aircraft of WW2 were monoplanes, as have been virtually all aircraft since. Thus the superiority of the design over the biplane was established, and biplanes have been relegated to specialized applications ever since.[3]

The high-wing of a de Havilland Canada Dash 8.

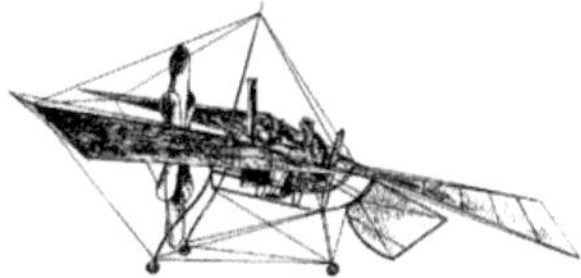

Félix du Temple's 1874 *Monoplane*.

A parasol wing Pietenpol Air Camper amateur-built aircraft.

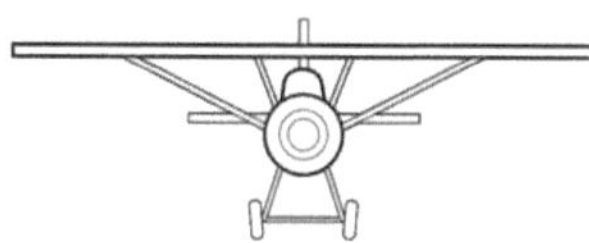

Schematic head-on illustration of a parasol wing

See also

- Biplane
- Triplane

References

[1] Gray, Carroll (undated). "Aleksandr Fyodorovich Mozhaiski" (http://www.flyingmachines.org/moz.html). . Retrieved 8 November 2011.

[2] King, *Windkiller*, p. 227.

[3] "Monoplane" (http://www.britannica.com/EBchecked/topic/390020/monoplane). *Encyclopædia Britannica, Encyclopædia Britannica Online*. Encyclopædia Britannica Inc. . Retrieved March 18, 2012.

[4] Compare, Trajan Vuia, (1872-1950) (http://www.ctie.monash.edu.au/hargrave/vuia.html)]

Sabena

Sabena

[[Image:Sabena Airlines Logo.svg		alt=]]

IATA	ICAO	Callsign
SN	SAB	SABENA

Founded	1923	
Ceased operations	2001	
Hubs	Brussels Airport	
Frequent-flyer program	Qualiflyer	
Airport lounge	Sabena lounge	
Alliance	Qualiflyer (now defunct)	
Fleet size	at time of bankruptcy: 87 (+15 orders)	
Destinations	At time of bankruptcy: 99	
Company slogan	*Enjoy Our Company*	
Headquarters	Brussels Airport Zaventem, Belgium	
Key people	At time of bankruptcy: Christophe Müller (CEO), Jean Louis Herremans (CFO)	
Website	Sabena.com [1]	

Sabena was the national airline of Belgium from 1923 to 2001, with its base at Brussels National Airport. After its bankruptcy in 2001, the newly formed SN Brussels Airlines took over part of Sabena's assets in February 2002, which then became Brussels Airlines. The airline had its corporate headquarters in the Sabena House on the grounds of Brussels Airport in Zaventem.[2]

History

1923–1939

SABENA is short for *Société Anonyme Belge d'Exploitation de la Navigation Aérienne* ('Belgian corporation for exploiting aerial navigation'). It began operations on 23 May 1923 as the national carrier of Belgium. It had been created by the Belgian government after its predecessor SNETA (*Syndicat national pour l'étude des transports aériens*) - formed in 1919 to pioneer commercial aviation in Belgium - ceased operations. The first commercial flight of Sabena was operated between Brussels and London (UK) on 1 July 1923 via Ostend. Services to Rotterdam (Netherlands) and Strasbourg (France) were launched on 1 April 1924. The Strasbourg service was extended to Basle (Switzerland) on 10 June 1924. Amsterdam (Netherlands) was added on 1 September 1924, and Hamburg (Germany) followed on 1 May 1929 via Antwerp, Düsseldorf, and Essen.

Belgian Congo

When Sabena was created, it was partly funded by Belgians in the Belgian Congo colony who lost their air service—an experimental passenger and cargo company (LARA) between Léopoldville, Lisala, and Stanleyville—a year earlier and expected the new Belgian national airline to fill this gap. On 12 February 1925, therefore, Sabena pioneered its long haul across Africa to Leopoldville, capital of the Belgian Congo. Throughout its history, Sabena had a long tradition of service to African destinations. For a long time, these were the only profitable routes served by the airline.

Sabena chose to use landplanes for its Congo operations and a program of aerodrome construction was initiated in the Congo. This was finished in 1926 and Sabena immediately began flights within the Belgian colony, the main route being Boma-Léopoldville-Élisabethville, a 2,288 km (1,422 mi) route over dense jungle. First, flights were operated with De Havilland DH.50 aircraft, although these were quickly replaced with the larger Handley Page W.8f airliners, which had three engines and offered ten seats.

By 1931 SABENA's fleet, including the aircraft used on the Congo network, totalled 43 aircraft. Its mainstay type was the Fokker F.VIIB with a lesser number of the smaller Fokker 7A and 14 Handley-Page types. It also used British Westland Wessex aircraft.

Sabena flew aircraft out to tropical Africa, Belgium's Congo colony, occasionally, but mostly these aircraft were shipped out. There was no direct flight yet between Belgium and its colony. As the 1930s progressed, SABENA cooperated with Air France and Deutsche Luft Hansa, which also had interests in routes to destinations across Africa.

SABENA's first long-haul flight to the Congo occurred on 12 February 1935 and took five and a half days, for which SABENA used a Fokker F-VII/3m aircraft. The following year, SABENA purchased the Savoia-Marchetti SM.73 airliner. With a speed of 300 km/h (200 mph), it reduced the journey time taken to only four days, and the SABENA service ran on alternate weeks to an Air Afrique service.

Expansion in Europe

In Europe, SABENA opened services to Copenhagen and Malmö in 1931. A route to Berlin was initiated in 1932. The mainstay pre-war airliner that SABENA used in Europe, was the successful Junkers Ju-52/3m airliner. The airline's pre-war routes covered almost 6,000 km within Europe. While the Brussels Haren airport was Sabena's main base, the company also operated services from other Belgian airports, and had a domestic network that was mainly used by businessmen who wanted to be in their coastal villas for the weekend.

In 1938, the airline purchased the new Savoia-Marchetti SM.83, a development of the S.M. 73 with a speed of 435 km/h (270 mph), although it flew services at a cruising speed of about 400 km/h (250 mph).

1939–1946

At the outbreak of the Second World War in 1939, SABENA's fleet totalled 18 aircraft. Its mainstay fleet type was the Savoia-Marchetti SM.73 airliner (it had 11 of the type) and the Junkers Ju-52/3m airliner (it had 5.) SABENA also had just taken delivery of two Douglas DC-3s.

During the war the airline managed to maintain its Belgian Congo routes, but all European services ceased.

1946–1960

After the Second World War in 1946, SABENA resumed operating a network of intra-European scheduled services. The fleet was initially consisted mainly of Douglas DC-3s (There were thousands of surplus C-47 Dakotas (the military variant of the DC-3) available to help airlines restart operations after the war.) The airline now flew under the name of **SABENA - Belgian World Airlines**.

On 18 September 1946, the SABENA DC-4 OO-CBG crashed near Gander International Airport (Newfoundland), killing 27 of the 44 people on board. Crash location [3][4] [5] [6]

Douglas DC-3 of Sabena in 1949

SABENA started its first transatlantic route to New York City on 4 June 1947, initially using unpressurised Douglas DC-4 airliners, which were augmented and later replaced by Douglas DC-6Bs. The DC-4s, followed by the DC-6s also restarted the airline's traditional route to the Belgian Congo. SABENA were first to introduce transatlantic schedules from the North of England, when the airline's DC-6B *OO-CTH* inaugurated its Brussels-Manchester-New York route on 28 October 1953.

Sabena Douglas DC-6B arriving at Manchester Airport from New York in 1955

The Convair 240 was introduced in 1949 to partially replace the DC-3's that had until then flown most European services. As of 1956, improved Convair 440 'Metropolitan' twins started to replace the Convair 240 twins and were used successfully well into the 1960s across European regional destinations.

In 1957, the long-haul Douglas DC-7C was introduced for long-haul routes but this plane would be supplanted after only three years by the jet age.

On 3 June 1954, a Sabena-operated DC-3 on a cargo flight from the UK to Yugoslavia was strafed by a Mikoyan-Gurevich MiG-15, killing the radio operator and wounding both the captain and engineer. Co-pilot Douglas Wilson managed to land in Austria but the plane suffered significant damage.[7]

For the 1958 world exposition in Brussels, Sabena leased two Lockheed Super Constellations from Seaboard World Airlines, using them mainly on transatlantic routes. In the same period, there were experiments with helicopter passenger service using Sikorsky S-58 aircraft from Brussels to Antwerp, Rotterdam, Eindhoven and the Paris heliport at Issy-les-Molineaux.

1960–1990

1960 saw the introduction of the new Boeing 707-320 intercontinental jet for long-haul trans-Atlantic flights to New York. SABENA was mainland Europe's first airline to operate a jet across the Atlantic (BOAC - now British Airways - had been flying jet transatlantic services using the de Havilland Comet 4 since 4 October 1958). Tragically, one of SABENA's aircraft became the first Boeing 707 to crash while in commercial service when Flight 548 crashed while preparing to land at Brussels on 15 February 1961. The United States Figure Skating Team was aboard the aircraft, en route from New York to Prague via Brussels to compete in a figure skating championship.

Sabena Douglas DC-6 at London Heathrow Airport in 1960.

Six Caravelle jetliners were introduced on all medium-haul routes in Europe from February 1961, being flown on most routes alongside the Convair 440s, until the early 1970s.

1961 also saw a major upheaval for SABENA in the Congo colony. Widespread rioting against Belgian colonials in the months leading up to, and after the independence of the Democratic Republic of the Congo, caused thousands of Belgians to flee the country. The Belgian government commandeered SABENA's entire long haul fleet to get the refugees back to Europe. Independence also meant the end of the impressive regional network of routes that the airline had built up in the Congo since 1924. When the new republic began its own airline, Air Congo, in June 1961, SABENA held 30 percent of that airline's shares.

Sabena Boeing 707-329 in April 1960, shortly after delivery.

The Douglas DC-6B aircraft remained in use with SABENA in the mid 1960s although they were no longer used on the airline's main routes; the Boeing 707s and Caravelles became the mainstay types during this decade.

Boeing 727-100s were introduced on important European routes from 1967 in a unique colour scheme; the fin markings incorporated bare-metal rudder and white engine colours. The only other aircraft to have its own special markings was the Douglas DC-10.

At this time Fokker F27 Friendships entered service between regional Belgian airports and European destinations such as London Heathrow.

Sabena Sud Aviation Caravelle.

1971 saw the introduction of the Boeing 747-100 on transatlantic routes flying alongside the Boeing 707-320Cs. SABENA, like many other trans-Atlantic airlines was satisfied with the Boeing 707s, but for commercial reasons recognised it had to buy the new jumbo-jets for its prestige services, notably New York JFK and as of the mid-seventies,

A Sabena Boeing 747-100 seen in 1976.

Chicago O'Hare. SABENA purchased only two first generation jumbo-jets and continued to fly the 707 into the late 1970s.

As of 1973, the Boeing 727s on the European network were replaced by the Boeing 737-200.

The Douglas DC-10-30 entered service in 1974. In total, SABENA purchased five of these convertible (Passengers and/or freight) wide-body jets.

In 1984 Airbus A310s were introduced on routes that had high passenger-density. This aircraft type also introduced a modernisation of the 1973 SABENA livery, in which a lighter blue was used and the titles on the fuselage were in a more modern style.

In June 1986 the first of two Boeing 747-300 aircraft joined the fleet, eventually replacing the older 747-100.

A Sabena Airbus A310 seen in 1985.

In an advertisement in National Geographic June 1982 (volume 161, no. 6) Sabena was priding itself with a superior inflight service. "And unlike most European airlines Sabena still offers First Class service within Europe and all the way through." This advertisement also states that "Sabena flies to 76 destinations on 4 continents" and uses the slogan "belgian world airlines" (written with no capital letters).

In 1989, Sabena invited Belgian fashion designer Olivier Strelli to create a new range of uniforms for their cabin crew.

1990–1995

A new name, **Sabena World Airlines**, and colours were introduced for the 1990s. The new livery had an overall white colour and the white circle tail logo in blue on the fin. A large 'sabena' title covered the fuselage in a light blue and the name "Belgian World Airlines" was at times barely visible although the title was also painted on the fuselage in small clear letters. The 1990s saw further fleet type renewal; the DC-10-30s were replaced with twin-engined Airbus A330 and four-engined Airbus A340.

A Sabena Boeing 737 at Dublin Airport in 1995.

After the liberalisation of the airline industry, it became apparent that Sabena had little chances to survive on its own in this very competitive market. The Belgian government, the main shareholder of the company, began searching for a fit partner.

Sabena remained in a catastrophic financial state. Year after year, the Belgian government had to fill the losses. The government was however prevented from funding due to EU anti-state aid rules. Consecutive governments charged themselves with looking for a fit partner for the airline.

In 1993, Air France purchased a large minority stake in Sabena, which it sold soon after. In 1995, Swissair purchased a 49 percent stake in SABENA.

In 1993 Sabena's head office was in the Air Terminus building on Rue Cardinal Mercier in Brussels.[8]

1995–2001

In March and April 1998 two McDonnell-Douglas MD-11 aircraft joined the fleet and long-haul destinations as Newark, Montreal and São Paulo were (re)introduced.

1999 saw new colours introduced in the SABENA fleet, beginning with an Airbus A340. One of the latest fleet types that SABENA has introduced, right after the A321 and A320 is the A319 which saw service in 2000. These new planes were part of a record-order of 36 Airbuses, imposed on SABENA when under Swissair management.

After an airline recession and the effects on the airline industry of the September 11 attacks in 2001 all airlines that flew across the Atlantic suffered badly.

A Sabena Boeing 737-500 in 2000.

Swissair had pledged to invest millions in SABENA, but failed to do so, partly because the airline had financial problems itself, having declared bankruptcy one month prior. SABENA operated its final flight on 7 November 2001. The company filed for legal protection against its creditors on 3 October, and went into liquidation on 6 November 2001. Fred Chaffart, Chairman of the Board of Directors of Sabena, read a declaration on this day to explain the decision. A group of investors managed to take over Delta Air Transport, one of SABENA's subsidiaries, and transformed it into SN Brussels Airlines.

Sabena BAe 146 at Birmingham, 2001

November 7, 2001 was the final day of operations for Sabena. Flight 690 from Abidjan to Brussels via Cotonou was the last Sabena flight to land in Brussels. An Airbus A340-300 registered as OO-SCZ operated the flight.

The Belgian Parliament formed a committee to investigate the reasons behind the bankruptcy of SABENA and the involvement of Switzerland's flag carrier. At the same time, the company's administrator investigated possible legal steps against Swissair, and its successor Swiss International Airlines.

2001–present

In 2006, the Belgian government, a former major shareholder of SABENA, filed criminal charges against the former Swissair management. The former Swissair management was condemned by the judges.[9]

On 16 January 2007 the Belgian - Flemish news program Terzake reported that during the nineties, several members of the board were paid large sums illegally through a SABENA affiliate in Bermuda. When Paul Reutlinger became the CEO of the company, he stopped the illegal payments. The program goes on to state that this might be the explanation why the Belgian board members remained quiet when it became apparent that Swissair was exploiting SABENA and, eventually, drove the company into bankruptcy.

SN Brussels Airlines BAe 146 in the former Sabena livery at London Heathrow Airport in 2002.

On December 14, 2007, Georges Jaspis, a former Second World War pilot in No. 609 Squadron RAF and the SABENA pilot with the most flying hours (27,000) died. Captain Jaspis was the pilot who inaugurated the Manchester to New York service in October 1953 and who collected the first Sabena Boeing 707 and 747. He had escaped Belgium during the war and made his way to England where he joined the Royal Air Force. He was awarded the Distinguished Flying Cross. He was buried in Opprebais, a village south-east of Brussels on December 19, 2007. The Belgian military attended and a flypast of four F-16 jets in missing man formation was made.

Reasons for bankruptcy

The reasons for SABENA's bankruptcy are numerous. One of the direct causes was Swissair not living up to their contractual obligations and failing to inject necessary funds into the company. This was because at the time Swissair was having its own financial problems. During the so called "Hotel agreement", signed on July 17, 2001, Belgian prime minister Guy Verhofstadt met with Swissair boss Mario Corti, who agreed to inject €258 million into SABENA. Mr Corti had made a terrible mistake as the sum was never paid, due to Swissair's own financial problems. The purchase of 34 new Airbus planes, imposed by the Swiss, was a burden SABENA could not cope with.

After the bankruptcy, a parliamentary commission in Belgium was established to investigate the demise of the airline. The Belgian politicians got a part of the blame; Rik Daems, who, at the time, was Minister of Public Enterprises & Participations, Telecommunication and Middle Classes, received most criticism due to his lack of effort. Swissair itself went bankrupt in October of that year.

Accidents and incidents

- On 7 December 1934 at least two SABENA aircraft were destroyed in a hangar fire at Evere as a result of a crash of a military Fairey Fox biplane.

Flights to or from Europe

- On 10 December 1935, a Savoia-Marchetti S.73 (registered OO-AGN) **crashed** at Tatsfield, Surrey, United Kingdom with the loss of eleven lives.
- On 16 November 1937, a Junkers Ju 52 (registered OO-AUB) crashed near Ostend, Belgium, while landing, killing all 12 people on board.
- On 17 September 1946 at 01:47 local time, a Douglas C-47 (registered OO-AUR) crashed upon take-off from Haren Airfield, killing one crew member. The other two crew and four passengers on board the flight that had been bound for Croydon Airport survived the accident.[10]
- The next day, 18 September 1946, saw 27 people losing their lives when a SABENA Douglas DC-4 (OO-CBG) crashed 35 km short of Gander Airport, where the aircraft had been planned to land for a refueling stop on the flight from Brussels to New York. At the time of the accident (07:42 UTC), there was dense fog near the airport, and the pilot of the DC-4 had executed a flawed, too low approach, so that the plane hit the ground. There were 17 survivors (16 passengers and one crew).[11]

The crash site of the DC-4 in Newfoundland.

- The 19 passengers and three crew members on a flight from Brussels to London lost their lives on 2 March 1948, when the aircraft, a Douglas DC-3 registered OO-AWH, crashed at 21:14 local time upon approach of London Heathrow Airport in low visibility conditions.[12]
- On 18 December 1949 at approximately 20:30 local time, a C-47 (registered OO-AUQ) crashed near Aulnay-sous-Bois, France, killing the four passengers and four crew on board. The aircraft had just left Paris – Le Bourget Airport for a flight to Brussels, when a wing malfunction was encountered.[13]
- On 14 October 1953 at ca. 15:20 local time, a Convair CV-240 (registered OO-AWQ) crashed near Kelsterbach, West Germany, killing the 40 passengers and four crew that had been on the flight from Frankfurt to Brussels.

Engine power had been lost upon take-off from Frankfurt Airport, making the aircraft impossible to control.[14]

- One passenger on board a flight from Brussels to Zurich was killed on 19 December of the same year, when the aircraft (a CV-240 registered OO-AWO) hit the ground 2.5 km short of the runway threshold of Kloten Airport at 18:55 local time. In low visibility conditions, the pilot had executed a flawed approach, which let the aircraft descend below the glidepath. The other 39 passengers and three crew members survived the accident.[15]

- On June 1954, a C-47 (registered OO-CBY) was attacked by a fighter aircraft near Maribor, Yugoslavia. The aircraft, that had been on a cargo flight from Blackbushe Airport to Belgrade, could be kept airborne, and a forced landing at Graz Airport was carried out, in which it ran off the runway. In the incident, one out of the four people on board was killed.[16]

- On 13 February 1955, the pilots of a SABENA flight from Brussels to Rome lost orientation when approaching Ciampino Airport, resulting in the aircraft involved, a Douglas DC-6 registered OO-SDB, crashing into the slope of Monte Terminillo at 18:53 local time, killing the 21 passengers and eight crew on board.[17]

- The disaster of **Flight 548** with its 73 casualties marked the worst accident in the history of SABENA. It happened on 15 February 1961 at 09:05 UTC, when the aircraft (a Boeing 707 registered OO-SJB) crashed at Brussels Airport following a flight from New York City.[18] Among the dead was the entire American delegation to the 1961 World Figure Skating Championships, slated to be held in Prague; the competition was canceled in the aftermath.

- On 13 July 1968, a cargo-configured 707 (registered OO-SJK) crashed upon approach of Lagos Airport on a flight from Brussels, killing the seven occupants. It was determined that the aircraft had been descending too low, so that it had struck trees.[19]

- A DC-3 registered OO-AUX (which SABENA had leased from Delta Air Transport) was damaged beyond repair in a ground accident at Amsterdam Airport Schiphol on 9 May 1970. The pilots began to taxi the aircraft even though they had not been cleared to do so, which resulted in the right propeller hitting an obstacle on the ground, and debris destroying the airliner beyond economic repair.[20]

- On 8 May 1972, **Flight 571** from Vienna to Tel Aviv with 101 people on board (a Boeing 707 registered OO-SJG) was hijacked by four members of the terrorist organization Black September, in order to secure the release of 315 detainees from Israeli prisons. At Ben-Gurion International Airport, two hijackers were shot and killed by the Israeli *Sayeret Matkal* special forces. One passenger died later of the wounds she had suffered in the shoot-out.[21]

- On 15 February 1978, a Boeing 707 (registered OO-SJE) overshot the runway at Los Rodeos Airport following a chartered holiday flight from Brussels with 189 passengers and seven crew on board. After all people could be evacuated, a fire erupted from spilled fuel, destroying the aircraft.[22]

- On 4 April of the same year at 18:07 local time, a Boeing 737-200 (registered OO-SDH) that had been on a crew training flight suffered a bird strike during landing practice at Charleroi Airport. The pilot instructor on board attempted to get the aircraft airborne again, but failed because the remaining runway length did not suffice, so that the runway was overshot, and the plane damaged beyond repair.[23]

- On 29 August 1998, **Flight 542** from New York to Brussels with 248 passengers and 11 crew members on board, which was operated using an Airbus A340-200 (registered OO-SCW), suffered a broken-off right landing gear upon landing at Brussels Airport, so that the plane veered off the runway. There were no notable injuries in the ensuing evacuation, and the aircraft could be repaired.[24]

- On 13 October 2000, **Flight 689** from Brussels to Abidjan was hijacked by a Nigerian national who was due to be deported. The Airbus A330-200 with 146 other passengers and 11 crew members on board was forced to land at Málaga Airport, were the perpetrator was overpowered by Spanish police forces.[25]

Flights in the Belgian colonies

- On 1 January 1943, a Junkers Ju 52 (registered OO-AUG) crashed near Bangui in then French Equatorial Africa.[26]
- On 25 March 1944, the Ju 52 registered OO-AGU was destroyed when it crashed at Costermansville Belgian Congo, now Bukavu[27]
- Only some days later, on 3 April, another aircraft of the same type (OO-AUF) crashed nearby.[28]
- On 14 December 1945, a Lockheed Model 18 Lodestar (registered OO-CAK) caught fire and was subsequently destroyed following a forced landing near Kouandé during a flight that had originated at Lagos.[29]
- On 7 January 1947, a Douglas C-47 (registered OO-CBO) crashed near Costermansville.[30]
- On 24 December 1947, a Lockheed Lodestar (OO-CAR) experienced an engine failure shortly after take-off from an airfield near Mitwaba, then French Congo, and subsequently crashed, killing the five occupants on board.[31]
- On 12 May 1948, at 11:00 local time, a DC-4 (registered OO-CBE) crashed in a thunderstorm during a scheduled passenger flight from Leopoldville to Libenge, then Belgian Congo, killing the 24 passengers and seven crew members.[32]
- Another 13 people (ten passengers, three crew) were killed on 31 August of that year, when their aircraft, a C-47 registered OO-UBL, crashed near Elizabethville on a flight from Manono.[33]
- On 27 August 1949, a Douglas C-47 (registered OO-CBK) experienced a loss of engine power shortly after take-off from Leopoldville Airport for a flight to Elizabethville with 17 passengers and three crew on board. The three crew members and two out of the seventeen passengers on board died in the ensuing crash.[34]
- On 24 July 1951, the right engine of a cargo-configured DC-3 (registered OO-CBA) caught fire upon take-off from Gao Airfield and crashed, resulting in the loss of lives of the three persons on board.[35]
- On 4 February 1952, a C-47 (registered OO-CBA) with sixteen occupants (twelve passengers, four crew) crashed near Kikwit following the mid-air break-up of a propeller, which resulted in vital parts of the airliner being destroyed by debris[36] . The aircraft was en route for a scheduled flight from Costermansville to Leopoldville. There were no survivors.
- On 18 May 1958, a Douglas DC-7 (registered OO-SFA) suffered a problem with its leftmost engine, whilst on a flight from Lisbon to Leopoldville with 56 passengers and nine crew members. The pilots prepared for an emergency landing at Casablanca-Anfa Airport, but shortly before touch-down, a go-around was attempted, which resulted in a stall because of the missing engine power. The aircraft crashed into buildings and caught fire at 04:25 local time, from which only four passengers could be saved alive.[37]

Fleet

Sabena's fleet consisted of the following aircraft at the time of the bankruptcy in November 2001:

- 15 Airbus A319-100
- 6 Airbus A320-200
- 3 Airbus A321-200
- 6 Airbus A330-200
- 4 Airbus A330-300
- 2 Airbus A340-200
- 2 Airbus A340-300
- 6 Boeing 737-300
- 5 Boeing 737-500
- 2 McDonnell Douglas MD-11

Historical aircraft

- Ansaldo A.300 (from SNETA) 1923-
- Avro XIX 1949-
- Bell 47 1950-1962
- Blériot-SPAD S.33 (from SNETA) 1923-
- Boeing 707 1959
- Boeing 727
- Boeing 747
- Bristol Freighter 1957-1964 operated by Air Charter
- Caudron Goéland 1949-
- Convair 240 1949-
- Convair 440 1956-
- de Havilland DH.4 (from SNETA) 1923-
- de Havilland DH.9 (from SNETA) 1923-
- de Havilland DH.50 1925-1937
- de Havilland Dove 1947-1954
- de Havilland Dragon Rapide 1949-
- Douglas DC-3 1939-
- Douglas DC-4 1946-
- Douglas DC-6 1947-
- Douglas DC-7 1956-
- Douglas DC-10
- Farman Goliath 1923-
- Fokker F.II 1927-1934
- Fokker F.VII 1929-1945
- Handley Page W.8 1924-1935
- Junkers Ju 52/3M 1936-1946
- Lockheed 14 1942-1954
- Lockheed Super Constellation three leased in 1958
- Lockheed Lodestar 1941-1942
- Rumpler C.IV (from SNETA) 1923-
- Rumpler C.VI (from SNETA) 1923-
- SABCA S.2 1926-1933
- Savoia Marchetti S.73 1935-1940
- Savoia Marchetti S.83 1938-1940
- Sikorsky S-55 1953-1956
- Sikorsky S-58 1956-1963
- Sikorsky S-62 1960-1961
- Sud Alouette II 1957-
- Sud Caravelle 1961-
- Vertol 44 1958
- Westland Wessex 1930-1938

Destinations

Africa

Abidjan, Banjul, Bamako, Bujumbura, Casablanca, Conakry, Cotonou, Dakar, Douala, Entebbe, Freetown, Johannesburg, Kigali, Kinshasa, Lagos, Lomé, Luanda, Monrovia, Nairobi, Ouagadougou, Sal, Yaoundé.

Asia [38]

Bangkok, Beirut, Bombay, Chennai, Jakarta, Kuala Lumpur, Manila, Singapore, Tehran, Tel Aviv, Tokyo-Narita.

Europe

Ajaccio, Amsterdam, Athens, Barcelona, Basel, Belfast, Berlin-Tempelhof, Bilbao, Birmingham, Bordeaux, Bologna, Bremen, Bristol, Bucharest, Budapest, Catania, Copenhagen, Dublin, Düsseldorf, Edinburgh, Faro, Florence, Frankfurt, Geneva, Glasgow, Hamburg, Hanover, Helsinki, Istanbul, Leeds/Bradford, Lisbon, Ljubljana, Luxembourg, London-City, London-Gatwick, London-Heathrow, Madrid, Manchester, Milan-Linate, Milan-Malpensa, Moscow, Munich, Nantes, Napels, Newcastle, Nice, Nuremberg, Oslo-Gardermoen, Paris-Charles de Gaulle, Paris-Orly, Palma de Mallorca, Porto, Prague, Rome, Saint-Petersburg, Seville, Sheffield, Sofia, Stockholm-Arlanda, Strasbourg, Stuttgart, Turin, Valencia, Venice, Verona, Vienna, Warsaw, Zurich.

North America

Atlanta, Boston, Chicago-O'Hare, Cincinnati/Northern Kentucky, Dallas, Toronto-Pearson, Montreal-Dorval, Montreal-Mirabel, New York-JFK, Newark, Washington-Dulles

Latin America

Buenos Aires, Montevideo, São Paulo, Santiago de Chile, Guatemala-City

References

Notes

[1] http://web.archive.org/web/*/http://www.sabena.com
[2] Von Schreiber, Sylvia. " Organisierte Pleite (http://www.spiegel.de/spiegel/print/d-20849245.html)." *Der Spiegel*. 26 November 2001. "Wenige Stunden vorher geschah noch weit Merkwürdigeres: Polizisten der Brüsseler "Aufspürungsbrigade 4" drangen in die Privatwohnungen von vier Managern und in das Firmengebäude Sabena House am Flughafen Zaventem ein."
[3] http://stable.toolserver.org/geohack/geohack.php?params=48.6920_N_54.9224_W_type:landmark
[4] http://www.zianet.com/tmorris/ganderrescue.html
[5] http://uscgaviationhistory.aoptero.org/images/RESCUE%20IN%20NEWFOUNDLAND.pdf
[6] http://uscgaviationhistory.aoptero.org/images/Ten%20feet%20Tall.pdf
[7] Aviation Safety Network (http://aviation-safety.net/database/record.php?id=19540603-0)
[8] "World Airline Directory." *Flight International*. 24–30 March 1993. 119 (http://www.flightglobal.com/pdfarchive/view/1993/1993 - 0642.html?search=Scanair).
[9] Sabena finally gets justice - the judges felt that the demise of Sabena was a consequence of non-compliance by Swissair contractual obligations - DBNet report January 2011 accessed 26 December 2011 (http://www.dhnet.be/infos/societe/article/340682/ sabena-obtient-enfin-justice.html)
[10] 17 September 1946 accident at the Aviation Safety Network (http://aviation-safety.net/database/record.php?id=19460917-0)
[11] ASN Aircraft accident Douglas DC-4-1009 OO-CBG Gander, NF (http://aviation-safety.net/database/record.php?id=19460918-0)
[12] March 1948 crash near Heathrow at the Aviation Safety Network (http://aviation-safety.net/database/record.php?id=19480302-0)
[13] December 1949 crash at the Aviation Safety Network (http://aviation-safety.net/database/record.php?id=19491218-0)
[14] October 1953 crash at the Aviation Safety Network (http://aviation-safety.net/database/record.php?id=19531014-0)
[15] December 1953 crash at the Aviation Safety Network (http://aviation-safety.net/database/record.php?id=19531219-0)
[16] 1954 military occurrence at the Aviation Safety Network (http://aviation-safety.net/database/record.php?id=19540603-0)
[17] 1954 crash at the Aviation Safety Network (http://aviation-safety.net/database/record.php?id=19550213-0)

[18] Flight 548 at the Aviation Safety Network (http://aviation-safety.net/database/record.php?id=19610215-3)

[19] 1968 crash at the Aviation Safety Network (http://aviation-safety.net/database/record.php?id=19680713-0)

[20] 1970 incident at the Aviation Safety Network (http://aviation-safety.net/database/record.php?id=19700509-2)

[21] 1970 hijacking at the Aviation Safety Network (http://aviation-safety.net/database/record.php?id=19720508-0)

[22] February 1978 incident at the Aviation Safety Network (http://aviation-safety.net/database/record.php?id=19780215-1)

[23] April 1978 incident at the Aviation Safety Network (http://aviation-safety.net/database/record.php?id=19780404-1)

[24] 1998 incident at the Aviation Safety Network (http://aviation-safety.net/database/record.php?id=19980829-1)

[25] 2000 hijacking at the Aviation Safety Network (http://aviation-safety.net/database/record.php?id=20001013-0)

[26] 1943 crash at the Aviation Safety Network (http://aviation-safety.net/database/record.php?id=19430101-0)

[27] March 1944 crash at the Aviation Safety Network (http://aviation-safety.net/database/record.php?id=19440325-2)

[28] April 1944 accident at the Aviation Safety Network (http://aviation-safety.net/database/record.php?id=19440403-3)

[29] 1945 incident at the Aviation Safety Network (http://aviation-safety.net/database/record.php?id=19451214-0)

[30] January 1947 crash at the Aviation Safety Network (http://aviation-safety.net/database/record.php?id=19470107-0)

[31] December 1947 crash at the Aviation Safety Network (http://aviation-safety.net/database/record.php?id=19471224-0)

[32] May 1948 crash at the Aviation Safety Network (http://aviation-safety.net/database/record.php?id=19480512-0)

[33] August 1948 crash at the Aviation Safety Network (http://aviation-safety.net/database/record.php?id=19480831-0)

[34] August 1949 crash at the Aviation Safety Network (http://aviation-safety.net/database/record.php?id=19490827-0)

[35] 1951 accident at the Aviation Safety Network (http://aviation-safety.net/database/record.php?id=19510724-0)

[36] 1952 crash at the Aviation Safety Network (http://aviation-safety.net/database/record.php?id=19520204-0)

[37] 1958 crash at the Aviation Safety Network (http://aviation-safety.net/database/record.php?id=19580518-0)

[38] Airline Route, taken from OAG. 1974: SABENA International Network (http://airlineroute.net/2008/12/22/sn-s74intl/) 22 December 2008.

External links

- Sabena (http://web.archive.org/web/*/http://www.sabena.com) (Archive)
- Sabena India (http://web.archive.org/web/*/sabenaindia.com) (Archive)
- Sabena Technics (http://www.sabenatechnics.com)
- Sabena Flight Academy (http://www.sfa.be)
- Sabeniens (http://www.sabeniens.com)
- Official Virtual Airline (http://www.svagroup.org/sabena/)
- Gumbel, Peter. " The Last Days of Sabena (http://www.time.com/time/magazine/article/0,9171,366278,00. html)." *TIME*. Sunday October 20, 2002.

Belgian_Congo

<table>
<tr><td colspan="2" align="center">Belgian Congo
Congo belge (French)
Belgisch-Kongo (Dutch)</td></tr>
<tr><td colspan="2" align="center">Belgian colony</td></tr>
<tr><td colspan="2" align="center">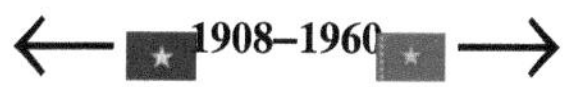1908–1908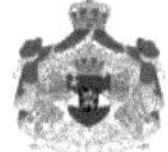</td></tr>
<tr><td colspan="2" align="center">Flag   Coat of arms</td></tr>
<tr><td colspan="2" align="center">Motto
Travail et Progrès
("Work and Progress")</td></tr>
<tr><td colspan="2" align="center">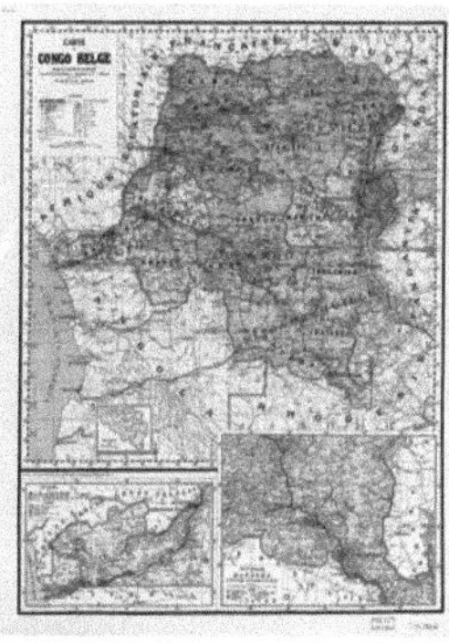
The Belgian Congo</td></tr>
</table>

Capital	Léopoldville/Leopoldstad
Language(s)	French (de facto official)[1] Dutch (majority of whites)[2] more than 200 indigenous languages
Political structure	Colony
King of the Belgians	
- 1908–09	Léopold II
- 1909–34	Albert I
- 1934–51	Léopold III
- 1951–60	Baudouin I
Governor-general	

- 1908–10	Théophile Wahis
- 1946–51	Eugène Jungers
- 1958–60	Henri Cornelis
History	
- Established	15 November 1908
- Independence	30 June 1960
- Secessions[1]	July–August 1960
Area	
- 1960	2344858 km^2 (905355 sq mi)
Population	
- 1960 est.	16610000
Density	7.1 /km^2 (18.3 /sq mi)
Currency	Congolese franc
[1] Secession of Katanga on 11 July and South Kasai on 8 August 1960	

The **Belgian Congo** (French: *Congo Belge*; Dutch: *Belgisch-Kongo*) was the formal title of present-day Democratic Republic of the Congo (DRC) between King Leopold II's formal relinquishment of his personal control over the state to Belgium on 15 November 1908, and Congolese independence on 30 June 1960.[3]

Congo Free State, 1884–1908

Leopold II, King of the Belgians and *de facto* owner of the Congo Free State from 1885 to 1908.

Until the later part of the 19th century, the Europeans had not yet ventured into the Congo. The rainforest, swamps and accompanying malaria, and other diseases, such as sleeping sickness, made it a difficult environment for European exploration and exploitation. In 1876, King Léopold II of the Belgians organized the International African Association with the cooperation of the leading African explorers and the support of several European governments for the promotion of African exploration and colonization. After Henry Morton Stanley explored the region, a journey that ended in 1878, Leopold courted the explorer and hired him to help establish Leopold's interests in the region.[4] Léopold II had been keen to acquire a colony for Belgium even before he ascended to the throne in 1865. He was convinced that the acquisition of a colony would bestow international prestige on his relatively young and small home country and that it might provide a steady source of income. Belgium was not greatly interested in its monarch's dreams of empire-building. Ambitious and stubborn, Léopold II decided to pursue the matter on his own account.

European rivalry in Central Africa led to diplomatic tensions, in particular with regard to the largely unclaimed Congo river basin. In November 1884, Otto von Bismarck convened a 14-nation conference (the Berlin Conference) to find a peaceful resolution to the Congo crisis. After three months of negotiation on 5 February 1885, the Berlin Conference reached agreement. While it did not formally approve or disapprove the territorial claims of the European powers in Central Africa, it did agree on a set of rules to ensure a conflict-free partitioning of the region. Key among those were the recognition of the Congo basin as a free-trade zone, and the general acceptance of the principle that any territorial claim needed to be backed up by evidence of actual and durable occupation of that territory. In reality, Léopold II emerged triumphant from the Berlin Conference.[5] In a series of bilateral diplomatic agreements, France was given **unknown operator: u','unknown operator: u','unknown operator: u',' (unknown operator: u'strong'unknown operator: u','**sq mi) on the north bank of the Congo river (modern Republic of the Congo and the Central African Republic), Portugal **unknown operator: u','unknown operator: u','unknown operator: u',' (unknown operator: u'strong'unknown operator: u','**sq mi) to the south (part of modern Angola), and Léopold's wholly owned, single-shareholder "philanthropic" organization received the balance: **unknown operator: u','unknown operator: u','unknown operator: u',' (unknown operator: u'strong'unknown operator: u','**sq mi), to be constituted as the Congo Free State.

The Congo Free State was a corporate state privately controlled by Léopold II, King of the Belgians through a dummy non-governmental organization, the *Association Internationale Africaine*. Léopold was the sole shareholder and chairman. The state included the entire area of the present Democratic Republic of the Congo and existed from 1885 to 1908, when it was annexed by the government of Belgium. Initially, the occupation and exploration of the immense territory of the Congo Free State proved a heavy burden on the monarch's purse. Twice, state bankruptcy was avoided by the Belgian state granting Léopold II emergency loans. In the 1890s, the tide turned dramatically. Through the forced exploitation of rubber, copper, and other minerals in the upper Lualaba River basin, together with the global rubber boom, huge surpluses were generated. Léopold II used part of this new wealth for the embellishment of his native country: the Royal Galleries in Ostend, the Palace of the Colonies in Tervuren, or the triumphal arch in Brussels were funded from the profits generated by the Congo. It soon became clear that these profits were generated on the back of brutal mistreatment of the local people and plunder of the Congo's natural resources.

1894 Congo Free State stamp.

Thus, under Léopold II's administration, the Congo Free State became the site of one of the worst man-made humanitarian disasters of the turn of the 20th century. The report of the British Consul Roger Casement, published in early 1904, was an irrefutable indictment of the "rubber system": "... the drowsy, unsupervised machine of coercion which wore out the people and the land".[6] In the absence of a census (the first was made in 1924), it is difficult to quantify the population loss of the period, but it must have been very high. According to Roger Casement's report, depopulation was caused mainly by four causes: "indiscriminate war", starvation, reduction of births, and tropical diseases. Adam Hochschild argues that roughly 10 million perished.[7] The human suffering inflicted by the rapacious exploitation of the colony was immense.

The European and American press exposed the conditions in the Congo Free State to the public in the early 1900s. In 1904, Léopold II was forced to allow an international parliamentary commission of inquiry entry to the Congo Free State. The report of the commission (1905) confirmed most of the charges formulated by Edmund Morel and Roger Casement, but also by Protestant and Roman Catholic missionaries.[8] By 1908, public pressure and diplomatic maneuvers led to the end of Léopold II's rule and to the annexation of the Congo as a colony of Belgium, known as the Belgian Congo.

Belgian colony, 1908–1960

Former residence of the Governor-general of the
Belgian Congo in Boma (picture 2008).

Steam Boat arriving at Shinkakasa (Congo
River), 1912.

Belgians residing in the Belgian Congo, 1900–1959		
Year	Pop.	±%
1900	1187	—
1910	1928	+62.4%
1920	3615	+87.5%
1930	17676	+389.0%
1939	17536	−0.8%
1950	39006	+122.4%
1955	69813	+79.0%
1959	88913	+27.4%
Source: [9]		

On 18 October 1908, the Belgian parliament voted in favor of annexing the Congo as a Belgian colony. This was only after King Léopold II had finally given up any hope to maintain a substantial part of the Congo Free State as separate crown property. The government of the Belgian Congo was arranged by the 1908 Colonial Charter.[10] Executive power rested with the Belgian Minister of Colonial Affairs, assisted by a Colonial Council (Conseil Colonial). Both resided in Brussels. The Belgian parliament exercised legislative authority over the Belgian Congo. The highest-ranking representative of the colonial administration in the Congo was the Governor-general. From 1886 until 1926, the Governor-general and his administration were posted in Boma, near the Congo River estuary. From 1926, the colonial capital moved to Léopoldville, some 300 km further upstream in the interior. Initially, the Belgian Congo was administratively divided into four provinces: Léopoldville (or: Congo-Kasaï), Equateur, Orientale and Katanga, each presided by a vice-Governor-general. An administrative reform in 1932 increased the number of provinces to six, while "demoting" the Vice-governor-generals to provincial Governors.

The territorial service was the true backbone of the colonial administration.[11] Each province was divided into a number of districts (24 in all), and each district into territories (some 120 in all). A territory was managed by a territorial administrator, assisted by one or more assistants. The territories were further subdivided into numerous "chiefdoms" (*chefferies*), at the head of which the Belgian administration appointed "traditional chiefs" (*chefs coutumiers*). The territories administered by one territorial administrator and a handful of assistants were often larger than a few Belgian provinces taken together (the whole Belgian Congo was nearly 80 times larger than the whole of Belgium). Nevertheless, the territorial administrator was expected to inspect his territory and to file detailed annual reports with the provincial administration. In terms of jurisdiction, two systems co-existed: a system of European courts and one of indigenous courts (*tribunaux indigènes*). These indigenous courts were presided over by the traditional chiefs, but had only limited powers and remained under the firm control of the colonial administration. Public order in the colony was maintained by the *Force Publique*, a locally recruited army under Belgian command. It was only in the 1950s that metropolitan troops—i.e., units of the regular Belgian army—were posted in the Belgian Congo (for instance in Kamina).

The colonial state—and in fact any authority exercised by whites in the Congo—was often referred to by the Congolese as *bula matari*. *Bula matari* ("break rocks") was one of the names originally given to Stanley, because of the dynamite he used to crush rocks when paving his way through the lower-Congo region.[12] The term *bula matari* came to signify the irresistible and compelling force of the colonial state.

When the Belgian government took over the administration from King Léopold II in 1908, the situation in the Congo improved in certain respects. The brutal exploitation and arbitrary use of violence, in which some of the concessionary companies had excelled, were curbed. The tragedy of "red rubber" was put to a stop. Article 3 of the new Colonial Charter of 18 October 1908 established that: "Nobody can be forced to work on behalf of and for the profit of companies or privates". In reality, forced labour, in differing forms and degrees, would not disappear entirely until the end of the colonial period.

The transition from the Congo Free State to the Belgian Congo was a break, but it was also marked by a large degree of continuity. The last Governor-general of the Congo Free State, Baron Wahis, remained in office in the Belgian Congo, and the majority of Léopold II's administration with him.[13] Opening up the Congo and its natural and mineral riches for the Belgian economy remained the main motive for colonial expansion, but all the same other priorities, such as healthcare and basic education, slowly gained in importance.

The Belgian Congo was directly involved in the two world wars. During WWI, an initial stand-off between the *Force Publique* and the German colonial army in German East-Africa (Tanganyika) turned into open warfare with a joint Anglo-Belgian invasion of German colonial territory. The *Force Publique* gained a notable victory when it marched into Tabora in September 1916. After the war, Belgium was rewarded for the participation of the *Force Publique* in the East African campaign with a League of Nations mandate over the former German colony of Ruanda-Urundi. During WWII, the Belgian Congo was a crucial source of income for the Belgian government in exile in London. The *Force Publique* again participated in the Allied campaigns in Africa. Congolese forces under the command of Belgian officers notably fought against the Italian colonial army in Ethiopia.[14]

Colonial economic policy

The economic development of the Congo was the colonizer's top priority. Under Belgian rule, two distinct periods of massive investment in the Congo's economic infrastructure stand out: the 1920s and the 1950s.[15]

Propaganda leaflet of the Belgian Ministry of Colonies, early 1920s.

After WWI, priority was given to mining (copper and cobalt in Katanga, diamond in Kasai, gold in Ituri) as well as to the transport infrastructure (rail lines Matadi-Léopoldville and Elisabethville-Port Francqui). To obtain the necessary capital, the colonial state gave the private companies, to a large extent, a free hand. This allowed, in particular, the Belgian Société Générale to build up an economic empire in the colony. Huge profits were generated and for a large part siphoned off to Europe in the form of dividends.[16] The necessary work force was recruited in the interior of the vast colony with the active support of the territorial administration. In many cases, this amounted to forced labour, as in many villages minimum quotas of "able-bodied workers" to be recruited were enforced. In this way, tens of thousands of workers were transferred from the interior to the sparsely populated copper belt in the south (Katanga) to work in the mines. In agriculture, too, the colonial state forced a drastic rationalisation of production. The so-called "vacant lands"—i.e., the land that was not directly used by the local tribes—fell to the state, which redistributed it to European companies, individual white landowners (*colons*) or the missions. This way an extensive plantation economy developed. Palm oil production in the Congo increased from 2,500 tons in 1914 to 9,000 tons in 1921 and 230,000 tons in 1957. Cotton production increased from 23,000 tons in 1932 to 127,000 in 1939.[17] After WWI the system of mandatory cultivation was introduced: Congolese peasants were forced to grow certain cash crops (cotton, coffee, groundnuts) destined for the European market. Territorial administrators and state agronomists had the task to supervise and if necessary sanction those peasants who evaded the hated mandatory cultivation.[18]

The mobilization of the African work force in the capitalist colonial economy played a crucial role in spreading the use of money in the Belgian Congo. The basic idea was that the development of the Congo had to be borne not by the Belgian taxpayers but by the Congolese themselves. The colonial state needed to be able to levy taxes in money on the Congolese, so it was important that they could earn money by selling their produce or their labour within the framework of the colonial economy.

The economic boom of the 1920s turned the Belgian Congo into one of the leading copper ore producers worldwide. In 1926 alone, the Union Miniére du Haut Katanga exported more than 80,000 tons of copper ore, a large part of which was processed in Hoboken in Belgium.[19] In 1928, King Albert I visited the Congo to inaugurate the so-called 'voie national' that linked the Katanga mining region via rail (up to Port Francqui) and river transport (from Port Francqui to Léopoldville) to the Atlantic port of Matadi. During the great depression of the 1930s, the export-based Belgian Congo economy was severely hit by the world crisis, because of the drop of international demand of raw materials and agricultural products (for example, the price of peanuts

Rwandese workers at the Kisanga-mine, Katanga, 1920s

fell from 1.25 francs to 25 cents). In some areas, as in the Katanga mining region, employment declined by 70% and in the whole country the exploitation of forced labour was diminished while many forced labourers returned to their villages.

After the occupation of Belgium by the Germans in May 1940, the Congo declared itself loyal to the Belgian government in exile in London to continue the war on the Allied side. During WWII, production increased drastically. After Malaysia fell to the Japanese, the Belgian Congo became a strategic supplier of rubber to the Allies. The Belgian Congo was one of the major exporters of uranium to the US during WWII (and the Cold War), particularly from the Shinkolobwe mine. The colony provided the uranium used in the atom bombs dropped on Hiroshima and Nagasaki in 1945.[14]

After WWII, the colonial state took on a much more active role in the economic and social development of the Belgian Congo. An ambitious ten-year plan was launched in 1949. It put a lot of emphasis on house building, energy supply and health care infrastructure. The ten-year plan ushered in a decade of strong economic growth, from which, for the first time, the Congolese began to benefit on a substantial scale. In 1953, the Congolese were granted the right to buy and sell private property in their own name. In the 1950s, a Congolese middle class, modest at first, but steadily growing, emerged in the main cities (Léopoldville, Elisabethville, Stanleyville, Luluabourg).

Railways (grey/black lines) and navigable waterways (purple lines) in the Belgian Congo, 1889-1960

Civilising mission

A key argument that was often invoked as a justification for colonialism in Africa was that of the so-called 'civilising mission' of the European nations.[20] This was no different with respect to the Belgian Congo. As elsewhere, this self-declared 'civilising mission' went hand in hand with the goal of economic exploitation and development. Conversion to Catholicism, basic western-style education and improved health care were objectives in their own right, but at the same time helped to integrate what was regarded a "primitive society" into the Western capitalist model, in which workers who were disciplined and healthy, and who had learned to read and write could be efficiently (and cheaply) put to work.

The development of education and health care in the Belgian Congo was impressive. The educational system was dominated by the Roman Catholic Church and, in some rare cases, Protestant churches, and the curricula reflected Christian and Western values. Even in 1948, 99.6% of educational facilities were controlled by Christian missions. Indigenous schooling was mainly religious and vocational. Children received basic education such as learning how to read, write and some mathematics. The Belgian Congo was one of the few African colonies in which local languages (Kikongo, Lingala, Tshiluba and Swahili) were taught at primary school. Even so, language policies and colonial domination often went hand in hand, as evidenced by the preference given to Lingala—a semi-artificial language spread through its common use in the Force Publique—over more local (but also more ancient) indigenous languages such as Lomongo and others.[21] In 1940 the schooling rates of children between 6 and 14 years old was

12%, reaching 37% in 1954, one of the highest rates in the whole of black Africa. Secondary and higher education for the indigenous population were not developed until relatively late in the colonial period. Black children, in small numbers, began to be admitted to European secondary schools from 1950 onward. The first university in the Belgian Congo, the Catholic University of Lovanium, near Léopoldville, opened its doors to black and white students in 1954.[22] In 1956 a state university was founded in Elisabethville.

Health care, too, was largely supported by the missions, although the colonial state took an increasing interest. Endemic diseases, such as sleeping sickness, were all but eliminated through large-scale and persistent campaigns.[23] The health care infrastructure expanded steadily throughout the colonial period, with a comparatively high availability of hospital beds relative to the population and with dispensaries set up in the most remote regions.

White nurses of the Union Minière du Haut Katanga and their Congolese assistants, Elisabethville, 1918.

There was a kind of "implicit apartheid", as there were curfews for Congolese city-dwellers and other such restrictions were commonplace. Though there were no specific laws (as in South Africa and the South of the United States at the time) barring blacks from entering the same establishments whites frequented, there was de facto segregation in most areas. For example, the city centres were reserved to white population only, while the blacks were organized in «cités indigènes» (ironically called 'le belge'). Hospitals, department stores and other facilities were often reserved for either whites or blacks. In the police, the blacks could not pass the rank of non-commissioned officer. The blacks in the cities could not leave their houses from 9 pm to 4 am. This type of segregation began to disappear gradually only in the 1950s.

King Albert I and Queen Elisabeth inspect the military camp of Leopoldville during their visit to the Belgian Congo, 1928.

The popular comic book Tintin in the Congo, first published in 1931, provides a good insight in the paternalistic or, as seen by some, racist views about 'primitive' Africa that prevailed at the time in Europe.

Because of the close interconnection between economic development and the 'civilising mission', and because in practice state officials, missionaries and the white executives of the private companies always lent each other a helping hand, the image has emerged that the Belgian Congo in reality was governed by a holy trinity of King-Church-Capital (or: the colonial state, the missions and the Société Générale de Belgique).

The ideology underpinning colonial policy was summed up in a catch-phrase used by Governor-general Pierre Ryckmans (1934–46): "Dominer pour servir" ("Dominate to serve").[24] The colonial government was keen to convey the image of a benevolent and conflict-free administration and of the Belgian Congo as a true model colony. In reality, no or very little attention was paid to the active emancipation of the Congolese. The coloniser alone knew what was good for the Congo. The local population was given no voice in the affairs of the state. It was only in the 1950s that this paternalistic attitude began to change. As from 1953, and even more so after the triumphant visit of King Baudouin to the colony in 1955, Governor-general Léon Pétillon (1952–58) actively favoured the creation of a "Belgian-Congolese community", in which blacks and whites were to be treated as equals.[25] In the 1950s, the most blatant discriminatory measures directed at the Congolese were hastily withdrawn (among these: the possibility to inflict corporal punishment by means of the feared *chicotte*—a fine whip of hippopotamus hide). In 1957, the first municipal elections open to black voters took place in a handful of the largest cities—Léopoldville, Elisabethville and Jadotville.

Resistance and voices of dissent

Congolese resistance against colonialism was widespread and took many different forms.[26] Armed resistance occurred sporadically and localized until roughly the end of the Second World War (e.g., revolt of the Pende in 1931, mutiny in Luluabourg 1944). From the end of the Second World War until the late 1950s, the so-called *pax belgica* prevailed. Until the end of colonial rule in 1960, passive forms of resistance and expressions of an anti-colonial sub-culture were manifold (e.g., Kimbanguism, after the prophet Simon Kimbangu, who was imprisoned by the Belgians).

Cathedral of the Jesuit mission in Kisantu, built in the 1930s.

Apart from active and passive resistance among the Congolese, the colonial regime over time also elicited internal criticism and dissent. Already in the 1920s, certain members of the Colonial Council in Brussels (among them Octave Louwers) voiced criticism regarding the often brutal recruitment methods employed by the major companies in the mining districts. The stagnation of population growth in many districts—in spite of spectacular successes in the fight against endemic diseases such as sleeping sickness—was another cause for concern. Low birth rates in the countryside and the depopulation of certain areas were typically attributed to the disruption of traditional community life as a result of forced labour migration and mandatory cultivation.[27] Many missionaries who were in daily contact with Congolese villagers, took their plight at heart and sometimes intervened on their behalf with the colonial administration (for instance in land property questions).

The missions and certain territorial administrators also played an important role in the study and preservation of Congolese cultural and linguistic traditions and artefacts. One example among many is that of Father Gustaaf Hulstaert (1900–1990) who in 1937 created the periodical 'Aequatoria' devoted to the linguistic, ethnographic and historical study of the Mongo-people of the central Congo basin.[28] The colonial state itself took an interest in the cultural and scientific study of the Congo, particularly after the Second World War through the creation of the *Institut pour la Recherche Scientifique en Afrique Centrale* (IRSAC, 1948).

Towards independence: 1945-1960

In the early 1950s, political emancipation of the Congolese elites, let alone of the masses, seemed like a far cry. Nonetheless, it was clear that the Congo could not forever remain immune from the rapid changes that, after the Second World War, profoundly affected colonialism around the world. The independence of the British, French and Dutch colonies in Asia shortly after 1945 had little immediate impact in the Congo, but in the United Nations pressure on Belgium (as on other colonial powers) was stepped up. Belgium had ratified article 73 of the United Nations Charter, which advocated self-determination, and both superpowers put pressure on Belgium to reform its Congo policy. However, the Belgian government tried to resist as best it could what it labeled 'interference' with its colonial policy.

All the same, it was clear to the colonial authorities that something needed to be done to ameliorate the situation of the Congolese. Since the 1940s, the colonial government had experimented in a very modest way with granting a limited elite of so-called évolués more civil rights, holding out the eventual prospect of a limited amount of political influence. To this end "deserving" Congolese could apply for a proof of "civil merit", or, one step up, 'immatriculation' (registration), i.e., official evidence of their assimilation with European civilisation. To acquire this status, the applicant had to fulfill strict conditions (monogamous matrimony, evidence of good behaviour, etc.) and submit to stringent controls (including house visits). This policy was a failure. By the mid-1950s, there were at best a few thousand Congolese

Palace of Justice in Elisabethville/Lubumbashi, built in Art Deco style. Photo 2010.

who had successfully obtained the civil merit diploma or been granted "immatriculation". The supposed benefits attached to it—including equal legal status with the white population—proved often more theory than reality and led to open frustration with the *évolués*. When Governor-General Pétillon began to speak about granting the native people more civil rights, even suffrage, to create what he termed a "Belgo-Congolese community", his ideas were met with indifference from Brussels and often with open hostility from some of the Belgians in the Congo, who feared for their privileges.[29]

It became increasingly evident that the Belgian government lacked a strategic long-term vision in relation to the Congo. This was due partly to the fact that 'colonial affairs' did not generate much interest or political debate in Belgium, so long as the colony seemed to be thriving and calm. A notable exception was the young King Baudouin I of the Belgians, who had succeeded his father, Léopold III, under dramatic circumstances in 1951, when Léopold was forced to abdicate because of his wartime role. Baudouin took a lively interest in the Congo. On his first state visit to the Belgian Congo in 1955, he was welcomed enthusiastically by cheering crowds of whites and blacks alike, as captured in André Cauvin's documentary film, *Bwana Kitoko*.[30] Foreign observers, such as the international correspondent of *The Manchester Guardian*, remarked that Belgian paternalism "seemed to work", and contrasted Belgium's seemingly loyal and enthusiastic colonial subjects with the restless French and British colonies. On the occasion of his visit, King Baudouin openly endorsed the Governor-General's vision of a "Belgo-Congolese community"; but, in practice, this idea progressed slowly. At the same time, divisive ideological and linguistic issues in Belgium, which heretofore had been successfully kept out of the colony's affairs, now began to make themselves felt in the Congo as well. These included the rise of unionism among workers, the call for public (state) schools to break the missions' monopoly on education, and the call for equal treatment in the colony of both national languages: French and Dutch. Until then, French had been promoted as the unique colonial language. The Governor-general feared that such divisive issues would undermine the authority of the colonial government in the eyes of the Congolese, while also diverting attention from the more pressing need for true emancipation.

Political organisation

As a result of the inability of the colonial government to introduce radical and credible changes, the Congolese elites began to take matters more and more in their own hands by organising themselves socially and soon also politically. In fact, it can be argued that the seeds of Congo's post-independence woes were sown in the emergence in the 1950s of two markedly different forms of nationalism among the Congolese elites. The nationalist movement—to which the Belgian authorities, to some degree, turned a blind eye—promoted territorial nationalism wherein the Belgian Congo would become one politically united state after independence. In opposition to this was the ethno-religious and regional nationalism that took hold in the Bakongo territories of the west coast, Kasai, and Katanga. The first political organisations were of the latter type. ABAKO, founded in 1950 as the *Association culturelle des Bakongo* ("lower-Congo region") and headed by Joseph Kasa-Vubu, was initially a cultural association that soon turned political and, from the mid-1950s, became a vocal opponent of Belgian colonial rule. Additionally, the organization continued to serve as the major ethno-religious organization for the Bakongo and became closely intertwined with the Kimbanguist church which was extremely popular in the lower Congo.

In 1955, Belgian professor Antoine van Bilsen published a treatise called *Thirty Year Plan for the Political Emancipation of Belgian Africa*.[31] The timetable called for the gradual emancipation of the Congo over a 30-year period—the time Van Bilsen expected it would take to create an educated elite who could replace the Belgians in positions of power. The Belgian government and many of the *évolués* were suspicious of the plan—the former because it meant eventually giving up the Congo, and the latter because Belgium would still be ruling Congo for another three decades. A group of Catholic *évolués* responded positively to the plan with a moderate manifesto in a Congolese journal called *Conscience Africaine*, with their only point of disagreement being the amount of Congolese participation.[32]

In 1957, by way of experiment, the colonial government organised in three urban centres (Léopoldville, Elisabethville and Jadotville) the first municipal elections in which Congolese people were allowed to stand for office and cast their vote. Events in 1957-58 led to a sudden acceleration in the demands for political emancipation. This was, in part, influenced by developments outside the Congo, notably the independence of Ghana in 1957 and President De Gaulle's August 1958 visit to Brazzaville, the capital of the French Congo, on the other side of the River Congo, opposite Léopoldville, in which he promised France's African colonies the free choice between a continued association with France or full independence. The World Exhibition organised in Brussels in 1958 (Expo 58) proved another eye-opener for many Congolese leaders, who were allowed to travel to Belgium for the first time.[33] In 1958, the demands for independence radicalised quickly and gained momentum. A key role was played by the Mouvement National Congolais (MNC). First set up in 1956, the MNC established itself in October 1958 as a national political party that supported the idea of a unitary and centralised Congolese nation after

Patrice Lumumba.

independence. Its most influential leader was the charismatic Patrice Lumumba. In 1959, an internal split was precipitated by Albert Kalonji and other MNC leaders who favoured a more moderate political stance (the splinter group was deemed Mouvement National Congolais-Kalonji. Despite the organisational divergence of the party, Lumumba's leftist faction (now the Mouvement National Congolais-Lumumba) and the MNC collectively had established themselves as by far the most important and influential party in the Belgian Congo. Belgium vehemently opposed Lumumba's leftist views and had grave concerns about the status of their financial interests should Lumumba's MNC gain power.

"Lipanda"

In the winter of 1958-59, while the Belgian government was debating a programme to gradually extend the political emancipation of the Congolese population, it was overtaken by events. On 4 January 1959, a prohibited political manifestation organised in Léopoldville by ABAKO got out of hand. At once, the colonial capital was in the grip of heavy rioting. It took the authorities several days to restore order and, by the most conservative count, several hundred died. The eruption of violence sent a shockwave through the Congo and Belgium alike. On 13 January, King Baudouin solemnly declared in a radio address that Belgium would work towards the full independence of the Congo "without hesitation, but also without irresponsible rashness"". Without committing to a specific date for independence, the government of prime minister Gaston Eyskens had a multi-year transition period in mind during which provincial elections would take place in December 1959, national elections in 1960 or 1961, after which administrative and political responsibilities would be gradually transferred to the Congolese, in a process presumably to be completed towards the mid-1960s. On the ground the reality looked quite different.[34] Increasingly, the colonial administration saw itself confronted with non-cooperation (e.g., refusal to pay taxes). In some regions anarchy threatened.[35] At the same time it was clear that an important portion of the Belgian population in the Congo opposed the idea of independence and felt betrayed by Brussels. In those circumstances, and faced with a radicalisation of Congolese demands, the chances of a gradual and carefully planned transition towards independence dwindled rapidly. In 1959, King Baudouin made another visit to the Belgian Congo. The contrast with his 1955 visit could not have been greater. Upon his arrival in Léopoldville/Kinshasa he was pelted with rocks by blacks who were angry with the imprisonment of Patrice Lumumba, convicted because of incitement against the colonial government. Though Baudouin's reception in other cities was considerably better, the shouts of "Vive le roi !" were often followed by "Indépendance immédiate !" The Belgian government wanted to avoid at all cost being drawn into a futile and potentially very bloody colonial war, as had happened to France in Vietnam and Algeria or to the Netherlands in Indonesia. For that reason, it was all the more inclined to give in to the demands for immediate independence voiced ever more vocally by the Congolese leaders. It was hoped that somehow, and in spite of the lack of preparations (including the lack of an educated elite: there were only a handful of Congolese holding a university degree at that time), miraculously, things might work out (what came to be known as "le pari congolais"—the Congolese bet).

In January 1960, Congolese political leaders were invited to Brussels to participate in a round-table conference to discuss independence. Patrice Lumumba was discharged from prison for the occasion. The conference agreed surprisingly quickly to grant the Congolese practically all of their demands: a general election to be held in May 1960 and full independence—"Dipenda"—on 30 June 1960. This was in no small measure thanks to the strong united front put up by the Congolese delegation. The political maneuvering ahead of the elections resulted in the emergence of three political alliances: a coalition of the federalistic nationalists consisting of six separatist parties or organisations, two of which were ABAKO and the MNC - Kalonji, the centralist MNC-Lumumba, and finally that of the strong-man of Katanga, Moïse Tshombe, conscious of the economic vitality of its area and the business interests of the Mining Union (just like Kalonji with respect to the diamond exploitations in Kasaï). The parliamentary elections resulted in a divided political landscape, with both the regionalist factions—chief among them ABAKO—and the nationalist parties such as the MNC, doing well. As time until independence day was running out, a compromise arrangement was forced through, with Kasa-vubu becoming the first president of the Republic of the Congo and Lumumba its first head of government. As planned, scarcely five months earlier, the hand-over ceremony took place on 30 June 1960. The location was the new residence of the Governor-General of the Belgian Congo in Léopoldville (which had only been recently built—in itself an indication of how unexpectedly independence came to the Congo). The ceremony was overshadowed by a significant incident: in his speech, King Baudouin, rather inappropriately, praised the genius of his forefather King Léopold II, founder of the Congo Free State, and the blessings of Belgian colonial rule. Prime Minister Lumumba retorted with a vehement indictment of colonial oppression.

Scarcely one week after the handover of sovereignty, a rebellion broke out within the *Force Publique* against its officers, who were still predominantly Belgian. This was the signal for disturbances all over the Congo, mainly instigated by dissatisfied soldiers and radicalised youngsters. In many areas, violence specifically targeted European victims. Within weeks, the largest part of the more than 80,000 Belgians who were still working and living in the Congo were evacuated in all haste and often under traumatic circumstances by the Belgian Army and later by the United Nations intervention force.[36]

Belgian Congo after 1960

The rebellion that had started in Thyssville in the Bas-Congo in July 1960 quickly spread to the rest of the Congo.[37] In September 1960, President Kasa-Vubu declared prime minister Lumumba deposed from his functions and vice versa. The stalemate was ended with the arrest of Lumumba. In January 1961, he was flown to the rich mining province of Katanga, which by that time had declared a secession from Léopoldville under the leadership of Moïse Tshombe (with active Belgian support). Patrice Lumumba was brutally murdered (in 2002 Belgium officially apologised for its role in the elimination of Lumumba; the CIA too has been suspected of complicity).[38] A series of rebellions and separatist movements seemed to shatter the dream of a unitary Congolese state at its birth. Although independent, Belgian paratroopers intervened in the Congo on various occasions to protect and evacuate fellow citizens. The United Nations maintained a large peace-keeping operation in the Congo from late 1960 onward. The situation stabilised only in 1964–65, with the re-integration of the Katanga province and the end of the so-called Simba Rebellion in Stanleyville (province Orientale). Shortly after that army colonel Joseph Désiré Mobutu ended the political impasse by seizing power himself.

Mobutu enjoyed the support of the West, and in particular of the United States, because of his strong anti-communist stance. Initially his rule favoured consolidation and economic development (e.g., by building the Inga-dam that had been planned in the 1950s). In order to distance himself from the previous colonial regime, he launched a campaign of Congolese "authenticity". As a result the colonial place names were abandoned in 1966: Léopoldville became *Kinshasa*, Elisabethville *Lubumbashi*, Stanleyville *Kisangani*. During this period, the Congo maintained close economic and political ties with Belgium, although these were occasionally overshadowed by the financial issues that had remained unresolved after independence (the so-called "contentieux"), for instance the transfer of shares in the big mining companies that had been held directly by the colonial state.[39] In 1970, on the occasion of the tenth anniversary of independence, King Baudouin paid an official state visit to the Congo.

Mobutu's régime radicalised during the 1970s. The *Mouvement populaire de la Révolution* (MPR), of which Mobutu was the président-fondateur, firmly established one-party rule. Political repression increased considerably. Mobutu now renamed the Congo into the republic of Zaïre. The so-called "zaïrisation" of economic life in the mid-1970s led to an exodus of foreign workers and an unmitigated economic disaster. In the 1980s the Mobutu regime became a byword for mismanagement and corruption.[40] Relations with the former colonial power Belgium went through a series of ups and downs, reflecting a steady decline in the underlying economic, financial and political interests. After the end of the Cold War, Mobutu lost support in the West. As a result, in 1990, he decided to end the one-party system and dramatically announced a return to democracy, but subsequently dragged his feet and played out his opponents against one another to gain time. A bloody intervention of the Zaïrian Army against students on the Lubumbashi University Campus in May 1990 precipitated a break in diplomatic relations between Belgium and Zaïre. Pointedly, Mobutu was not invited to attend the funeral of King Baudouin in 1993, which he considered a grave personal affront. Finally, in 1997 Mobutu was chased from power by a rebel force headed by Laurent-Désiré Kabila, who declared himself president and renamed Zaïre into the Democratic Republic of the Congo. Assassinated in 2001, Laurent-Désiré Kabila was succeeded by his son Joseph Kabila, who in 2006 was confirmed as president through the first nation-wide free elections in the Congo since 1960. On 30 June - 2 July 2010, King Albert II of the Belgians and Yves Leterme, the Belgian Prime Minister, visited Kinshasa to attend the festivities marking the 50th anniversary of Congolese independence from Belgium.

Certain practices and traditions from the colonial period have survived into the independent Congolese state, such as a strong centralising and bureaucratic tendency, or the organisational structure of the education system and the judiciary. The influence of the Congo on Belgium has manifested itself mainly in economic terms: through the activities of the Union Minière (now Umicore), the development of a nonferrous metal industry, and the development of the Antwerp harbour and diamond industry. To this day, Brussels Airlines (successor of the former Sabena) has maintained a strong presence in the DRC. It is estimated that there currently (2010) remain more than 4,000 Belgians resident in the DRC, while the Congolese community in Belgium is at least 16,000 strong. The

"Matonge" quarter in Brussels (Porte de Namur) is the traditional meeting point of the Congolese community in Belgium.[41]

In popular culture

The Belgian Congo features prominently or as a backdrop in some great works of Western literature. Among those are:

- Joseph Conrad (1903) *Heart of Darkness*
- André Gide (1927), *Voyage au Congo*;
- Herge (1930), *Tintin in the Congo*;
- Evelyn Waugh (1931), *Remote people*;
- Kathryn Hulme (1956), *The Nun's Story (also 1959 film by Fred Zinnemann, starring Audrey Hepburn)*;
- Graham Greene (1959), *A burnt-out case*;
- Lieve Joris (1987), *Terug naar Congo*.
- Barbara Kingsolver (1998), *The Poisonwood Bible*

"Colonie belge" - usually depicting a black prisoner being flogged by a black guardian under the watchful eye of a white official — is a recurring theme in Congolese paintings by artists like Tshibumba Kanda-Matulu, C. Mutomobo and others.[42]

In the song "We Didn't Start the Fire" by Billy Joel, Belgian Congo is mentioned once in the second verse: "Belgians in the congo".

In "Short Memory," by Midnight Oil, the line "Belgians in the Congo" appears in the first verse.

Governors-General

- Baron Théophile Wahis (November 1908 – May 1912; originally appointed by Leopold II in 1900)
- Félix Alexandre Fuchs (May 1912–January 1916)
- Eugène Joseph Marie Henry (January 1916–January 1921)
- Maurice Eugène Auguste Lippens (January 1921–January 1923)
- Martin Joseph Marie René Rutten (January 1923–December 1927)
- Auguste Constant Tilkens (December 1927–September 1934)
- Pierre Marie Joseph Ryckmans (September 1934–July 1946)
- Eugène Jacques Pierre Louis Jungers (July 1946–January 1952)
- Léon Antoine Marie Pétillon (January 1952–July 1958)
- Henri Arthur Adolf Marie Christopher Cornelis (July 1958–June 1960)

See also

- Free Belgian Forces
- *Heart of Darkness*
- *Tintin in the Congo*
- University of Lovanium
- "We Didn't Start the Fire" (song by Billy Joel)
- King Leopold's Ghost
- Patrice Lumumba
- Lumumba (film), a biographical film directed by Raoul Peck
- List of colonial governors of the Congo Free State and Belgian Congo

References

[1] (French) République démocratique du Congo (http://www.tlfq.ulaval.ca/axl/afrique/czaire.htm), Laval University, Canada

[2] (Dutch) Vlamingen en Afrikanen—Vlamingen in Centraal Afrika (http://soc.kuleuven.be/arc/afrikaverteltd/?q=book/print/298), Faculteit Sociale Wetenschappen, Katholieke Universiteit Leuven, Belgium

[3] Georges Nzongola-Ntalaja (2002). *The Congo from Leopold to Kabila: A People's History*. Zed Books. ISBN 1-84277-053-5.

[4] Hochschild 61–67.

[5] Hochschild 84–87.

[6] Ascherson, Neal (1963), The King Incorporated, Leopold the Second and the Congo, London: Granta Books (ed. 1999), p. 250.

[7] Hochschild, adam (1998). *King Leopold's Ghost*.

[8] Stengers, Jean, "Le rôle de la commission d'enquête de 1904-1905 au Congo". In *Congo: Mythes et réalités*, Bruxelles: Editions Racine, 2005, pp. 159-179.

[9] Vanthemsche, Guy (2007), *La Belgique et le Congo*, Brussels: Editions Complexe, pp. 353–4.

[10] Senelle, R., and E. Clément (2009), *Léopold II et la Charte Coloniale*, Brussels: Editions Mols.

[11] A good overview in: Dembour, Marie-Bénédicte (2000), *Recalling the Belgian Congo, Conversations and Introspection*, New York: Berghahn Books, pp. 17–44.

[12] Likaka, Osumaka (2009), *Naming Colonialism, History and Collective Memory in the Congo, 1870–1960*, Madison: University of Wisconsin Press, p. 56.

[13] Stengers, Jean (2005), *Congo: Mythes et réalités*, Brussels: Editions Racine.

[14] McCrummen, Stephanie (2009-08-04). "Nearly Forgotten Forces of WWII". *The Washington Post*. Washington Post Foreign Service.

[15] Vanthemsche, Guy (2007), *La Belgique et le Congo*, Brussels: Editions Complexe.

[16] Buelens,Frans (2007), *Congo 1885-1960, Een financiëel-economische geschiedenis*, Berchem: EPO.

[17] Boahen, A. Adu (1990). *Africa Under Colonial Domination, 1880–1935*. p. 171.

[18] Likaka, Osumaka (1997), *Rural Society and Cotton in Colonial Zaire*, Madison: University of Wisconsin Press.

[19] Brion, René and Jean-Louis Moreau (2006), *De la Mine à Mars: la genèse d'Umicore*, Tielt: Lannoo.

[20] Young, Crawford (1994), *The African Colonial State in Comparative Perspective*, New Haven: Yale University Press, pp. 122–24 and 156–58.

[21] On this subject see for instance: Fabian, Johannes (1986), *Language and Colonial Power, The Appropriation of Swahili in the Former Belgian Congo 1880–1938*, Berkeley: University of California Press.

[22] Mantels, Ruben (2007), Geleerd in de tropen, Leuven, Congo en de wetenschap, 1885-1960, Leuven: Universitaire Pers, pp. 201-236.

[23] A critical assessment of the colonial obsession with sleeping sickness in: Lyons, Maryinez (1992), *The Colonial Disease, A social history of sleeping sickness in northern Zaire, 1900–1940*, Cambridge: Cambridge University Press.

[24] Vanderlinden, Jacques (1994), *Pierre Ryckmans 1891-1959, Coloniser dans l'honneur*, Brussels: De Boeck.

[25] Pétillon, L. A. M. (1967), *Témoignage et réflexions*, Brussels: Renaissance du Livre.

[26] Likaka, Osumaka (2009), *Naming Colonialism, History and Collective Memory in the Congo, 1870–1960*, Madison: University of Wisconsin Press.

[27] see for instance a lecture by Nancy Rose Hunt (2002), *Rewriting the Soul in Colonial Congo: Flemish Missionaries and Infertility*, Antwerp University: http://www.nias.knaw.nl/Ortelius2002.pdf.

[28] See: www.aequatoria.be

[29] Ndaywel è Nziem, Isidore (1998), *Histoire générale du Congo*, Paris-Brussels: De Boeck & Larcier, pp. 456–63.

[30] Raspoet, Erik (2005). *Bwana Kitoko en de koning van de Bakuba*. Meulenhoff/Manteau. ISBN 90-8542-020-2.

[31] Gerard-Libois, Jules (1989), "Vers l'Indépendance: une accélération imprévue", In *Congo-Zaïre*, Brussels: GRIP, pp. 43-56.

[32] Kalulambi Pongo, Martin (2009), "Le manifeste 'Conscience africaine: genèse, influences et réactions", In Tousignant, Nathalie (ed.), *Le manifeste Conscience africaine, 1956*, Brussels: Facultés Universitaires Saint-Louis, pp. 59-81.

[33] Aziza Etambala, Zana (2008), *De teloorgang van een modelkolonie, Belgisch Congo 1958-1960*, Leuven: Acco, pp. 105-110.

[34] Young, Crawford (1965), *Politics in the Congo, Decolonization and independence*, Princeton: Princeton University Press, pp. 140–161.

[35] Ryckmans, Geneviève (1995), *André Ryckmans, un territorial du Congo belge*. Paris. L'Harmattan, pp. 215-224.

[36] Verlinden, Peter (2002), *Weg uit Congo, Het drama van de kolonialen*, Leuven: Davidsfonds

[37] For an overview of developments in the Congo after 1960 see: O'Ballance, Edgar (2000), *The Congo-Zaire Experience, 1960–98*, Houndmills: MacMillan Press.

[38] An illuminating first-hand account of the CIA's activities in the Congo in 1960–61 in: Devlin, Larry (2008), *Chief of Station, Congo: Fighting the Cold War in a Hot Zone*, Cambridge: PublicAffairs

[39] Willame, Jean-Claude (1989), "Vingt-cinq ans de rélations belgo-zaïroises", In Congo-Zaïre, Brussels: GRIP, pp. 145-58.

[40] Wrong, Michela (2001), *Living on the brink of disaster in Mobutu's Congo, In the footsteps of Mr Kurtz*, New York: HarperCollins, pp. 195–200.

[41] Swyngedouw, Eva and Erik Swyngedouw (2009), "The Congolese Diaspora in Brussels and hybrid identity formation", In *Urban Research & Practice*, vol 2, 1, pp. 68-90.

[42] Fabian, Johannes (1996), *Remembering the Present, Painting and Popular History in Zaïre, Narrative and Paintings by Tshibumba Kanda Matulu*, Berkeley: University of California Press.

Books

- Hochschild, Adam (2005). *King Leopold's Ghost : A Story of Greed, Terror, and Heroism in Colonial Africa* (Mariner Books ed.). Boston: Houghton Mifflin Company. ISBN 978-0-618-00190-3.
- Ndaywel è Nziem, Isidore (1998). *Histoire générale du Congo*. Paris and Brussels: De Boeck & Larcier.
- Stengers, Jean (2005). *Congo, Mythes et réalités*. Brussels: Editions Racines.
- Van Reybrouck, David (2010). *Congo, Een geschiedenis*. Amsterdam: De Bezige Bij.

External links

- Belgian Congo (http://www.1911encyclopedia.org/Belgian_Congo_(addition)) article in *Encyclopædia Britannica* 1922 extension.
- Oasis Kodila Tedika and Francklin Kyayima Muteba. *Sources of growth in Democratic Republic of the Congo before independence. A cointegration analysis. Revue congolaise d'économie / Congo Economic Review.* Document de Travail / Working Paper. WP02/10 — July 2006. (http://congoeconomie.org/growth accounting drc version WP-DT2-2010.pdf)

Gnome-Rhône_Mistral_Major

<table>
<tr><td colspan="2" align="center">Mistral Major</td></tr>
<tr><td colspan="2">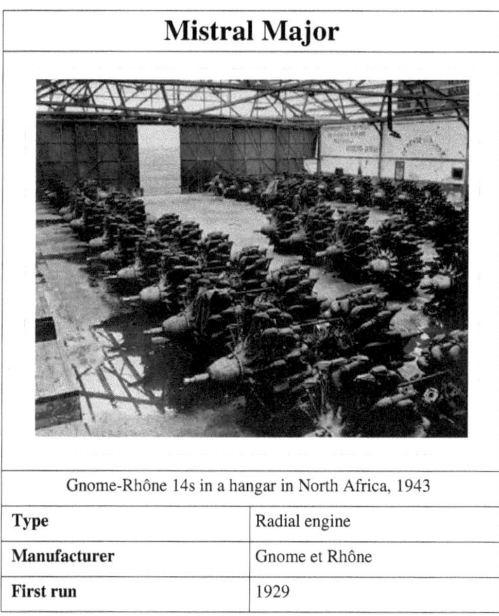</td></tr>
<tr><td colspan="2" align="center">Gnome-Rhône 14s in a hangar in North Africa, 1943</td></tr>
<tr><td>Type</td><td>Radial engine</td></tr>
<tr><td>Manufacturer</td><td>Gnome et Rhône</td></tr>
<tr><td>First run</td><td>1929</td></tr>
</table>

The **Gnome-Rhône 14K** *Mistral Major* was a 14-cylinder, two-row, air-cooled radial engine. It was Gnome-Rhône's major aircraft engine prior to World War II, and matured into a highly sought-after design that would see licensed production throughout Europe and Japan. Thousands of Mistral Major engines were produced, used on a wide variety of aircraft.

Design and development

In 1921 Gnome-Rhône purchased a license for the highly successful Bristol Jupiter engine and produced it until about 1930, alongside the smaller Bristol Titan. Starting in 1926, however, they used the basic design of the Titan to produce a family of new engines, the so-called "K series". These started with the 5K *Titan*, followed by the 7K *Titan Major* and 9K *Mistral*. By 1930, 6,000 of these engines had been delivered.

However, the aircraft industry at that time was rapidly evolving and producing much larger aircraft that demanded larger engines to power them. Gnome-Rhône responded by developing the 7K into a two-row version that became the 14K *Mistral Major*. The first test examples were running in 1929.

Licence built derivatives

- Manfred Weiss WM K.14
- Piaggio P.XI
- IAR 14K
- Tumansky M-87

Applications

- Amiot 143
- Aero A.102
- Bloch MB.200
- Bloch MB.210
- Breguet 274
- Breguet 460
- Breguet 521
- Dornier Do 17K
- Farman F.222
- Loire 46 C1
- PZL P.24
- PZL P.43
- Potez 62
- Potez 651

Aircraft powered by G-R 14K derivatives

- Aero A.102
- Breda Ba.65
- Breda Ba.88
- CANT Z.1007
- CANT Z.1011
- Caproni Ca.135
- Caproni Ca.161
- Heinkel He 70
- IAR 37
- IAR 80
- Ilyushin DB-3
- MÁVAG Héja
- Reggiane Re.2000
- Saab 17
- Savoia-Marchetti SM.79
- Savoia-Marchetti SM.84
- Sukhoi Su-2
- Weiss WM-21 Sólyom

Specifications (Gnome-Rhône 14Kdrs)

Data from [1]

General characteristics

- **Type:** Fourteen-cylinder two-row air-cooled radial engine
- **Bore:** 146 mm (5.75 in)
- **Stroke:** 165 mm (6.5 in)
- **Displacement:** 38.72 l (2,363 in³)
- **Diameter:** 1,296 mm (51.02 in)
- **Dry weight:** 540 kg (1,190 lb)

Components

- **Valvetrain:** Overhead valves
- **Supercharger:** Single-speed centrifugal type supercharger
- **Fuel system:** Stromberg carburetor
- **Fuel type:** 87 octane rating gasoline
- **Cooling system:** Air-cooled
- **Reduction gear:** 2:3

Performance

- **Power output:**

  - 743 kW (996 hp) at 2,390 rpm for takeoff
  - 821 kW (1,100 hp) at 2,390 rpm at 2,600 m (8,530 ft)
- **Specific power:** 21.23 kW/l (0.47 hp/in³)
- **Compression ratio:** 5.5:1
- **Specific fuel consumption:** 328 g/(kW•h) (0.54 lb/(hp•h))
- **Oil consumption:** 20 g/(kW•h) (0.53 oz/(hp•h))
- **Power-to-weight ratio:** 1.52 kW/kg (0.92 hp/lb)

See also

Related development

- IAR 14K
- Piaggio P.XI
- Tumansky M-87

Related lists

- List of aircraft engines

References

[1] Tsygulev (1939). *Aviacionnye motory voennykh vozdushnykh sil inostrannykh gosudarstv ([[Russian language\Russian* (http://base13. glasnet.ru/text/aviamotory/t.htm)]: Авиационные моторы военных воздушных сил иностранных государств)]. *Moscow: Gosudarstvennoe voennoe izdatelstvo Narkomata Oborony Soyuza SSR.* .

• Danel, Raymond and Cuny, Jean. *L'aviation française de bombardement et de renseignement 1918-1940* Docavia n°12, Editions Larivière

Gnome-Rhône_9K

9K	
Type	Radial engine
Manufacturer	Gnome et Rhône

The **Gnome-Rhône 9K** was a 9-cylinder 550 hp (405 kW) to 700 hp air-cooled radial engine, that started life as an enlarged Gnome-Rhône 7K with two extra cylinders. The Gnome-Rhône 7K itself was an enlarged version of the Gnome-Rhône 5K which was derived from a licensed version of the Bristol Titan . A redesign of the cylinders is indicated by the K suffix. The 9K was followed by the larger and more powerful 14-cylinder twin row Gnome-Rhône 14K.

The 9K was also produced in the Soviet Union as the M-75 at factory No. 29 in Zaporozhye. Only small numbers were built and it was dropped in favor of the M-25 a version of the Wright Cyclone and the M-85 a version of the 14K Mistral Major.

Specifications

General characteristics

- **Type:** Nine-cylinder air-cooled radial engine.
- **Bore:** 5.75 in (146 mm)
- **Stroke:** 6.5 in (165 mm)
- **Displacement:** 1,517in³ (24.9 Liters)

Components

- **Fuel system:** Carburetor
- **Cooling system:** Air-cooled

Applications

- IAR 15
- Breguet Calcutta
- Loire 70
- Morane Saulnier MS 225
- Wibault 365
- Savoia Marchetti S73 (license build Piaggio Stella P.IX)
- Savoia Marchetti SM81 (license build Piaggio Stella P.IX)

Radial_engine

The **radial engine** is a reciprocating type internal combustion engine configuration in which the cylinders point outward from a central crankshaft like the spokes on a wheel. Multiple rows of radial cylinders can be used for increased capacity. The radial engine was very commonly used in large aircraft engines before most large aircraft started using turbine engines.

Since the axes of the cylinders are coplanar, the connecting rods cannot be attached to the crankshaft as usual (Inline-four_engine). Instead, the pistons are connected to the crankshaft with a master-and-articulating-rod assembly. One piston, the uppermost one in the animation, has a master rod with a direct attachment to the crankshaft. The remaining pistons pin their connecting rods' attachments to rings around the edge of the master rod.

Four-stroke radials always have an odd number of cylinders per row, so that a consistent every-other-piston firing order can be maintained, providing smooth operation. For example, on a 5-cylinder engine the firing order is 1,3,5,2,4 and back to cylinder 1. Moreover, this always leaves a one-piston gap between the piston on its combustion stroke and the piston on compression. The active stroke directly helps compressing the next cylinder to fire, so making the motion more uniform. If an even number of cylinders was used, the equally timed firing cycle would not be feasible. [1] The protoype radial Zoche aero-diesels (below) have an even number of cylinders, either four or eight; but this is not problematic, because they are two-stroke engines, with twice the number of power strokes as a four-stroke engine.

The radial engine uses very few cams compared to other types. As usual for a four-stroke, the crankshaft takes two revolutions to complete the four strokes of each piston (intake, compression, combustion, exhaust). The camshaft is coaxial with the crankshaft and is typically geared with a

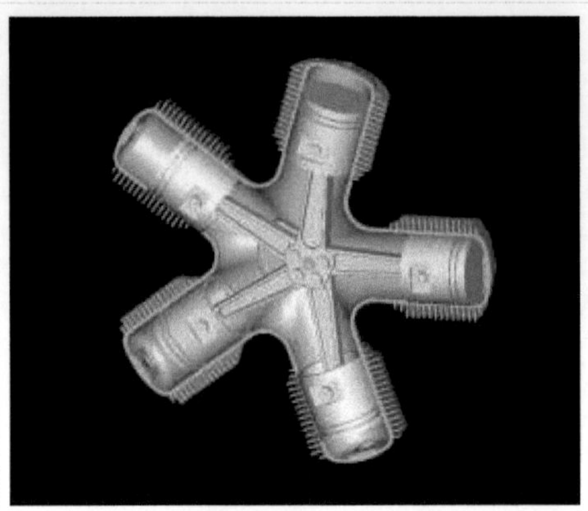

Radial engine in a cut-away view

Radial engine of a biplane

Master rod (upright), slaves and balances from a two-row, seven-cylinder Pratt & Whitney Twin Wasp

1:(1-n) transmission ratio, where n is the number of cylinders. As shown in the animation [2], the camshaft spins slower and in the opposite direction. The actual cams (cam lobes) are placed on two rows for the intake and exhaust. For the example, only 4 cams serve all 5 cylinders, whereas 10 would be required for a typical inline engine with the same number of cylinders.

Most radial engines use overhead poppet valves driven by pushrods and lifters on a cam plate which is concentric with the crankshaft, with a few smaller radials, like the five-cylinder Kinner B-5 and Russian Shvetsov M-11, using individual camshafts within the crankcase for each cylinder. A few engines utilize sleeve valves instead, like the very reliable 14-cylinder Bristol Hercules (built up to 1970 under licence in France by SNECMA) and the powerful 18-cylinder Bristol Centaurus.

History

C. M. Manly constructed a water-cooled five-cylinder radial engine in 1901, a conversion of one of Stephen Balzer's rotary engines, for Langley's *Aerodrome* aircraft. Manly's engine produced 52 hp (**unknown operator: u'strong'** kW) at 950 rpm.[3]

In 1903-04 Jacob Ellehammer used his experience constructing motorcycles to build the world's first air-cooled radial engine, a three-cylinder engine which he used as the basis for a more powerful five-cylinder model in 1907. This was installed in his triplane and made a number of short free-flight hops. During 1908-9, Ellehammer developed another engine, which had six cylinders arranged in two rows of three. His engines had a very good power-to-weight ratio, but his aircraft designs suffered from his lack of understanding of control. If he had concentrated on his engines, he might have become a successful manufacturer.[4]

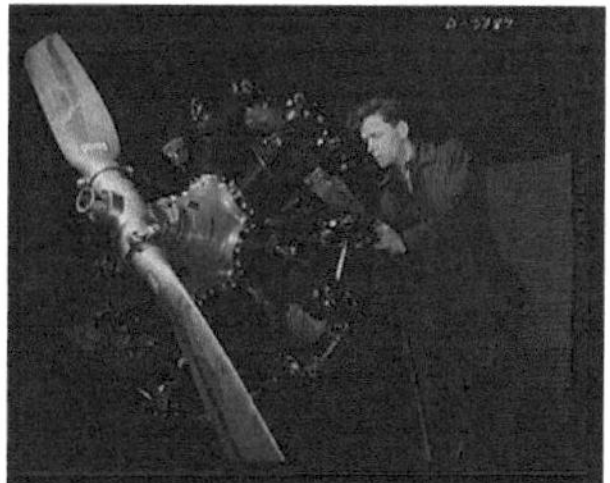

A Continental radial engine, 1944

Another early radial engine was the three-cylinder Anzani, originally built as a "semi-radial" W3 configuration design, one of which powered Louis Blériot's Blériot XI in his July 25, 1909 crossing of the English Channel. By 1914 Anzani had developed their range, their largest radial being a 20-cylinder engine of 200 hp (**unknown operator: u'strong'** kW), with its cylinders arranged in four groups of five.[3] One of the three-cylinder "fully radial", 120° cylinder angle Anzani powerplants still exists today, in fully running condition, in the nose of Old Rhinebeck Aerodrome's restored and flyable 1909 vintage Blériot XI. There is also another running Anzani at Brodhead airfield to go on a replica Blériot XI.

Pratt & Whitney R-1340 radial engine mounted in Sikorsky H-19 helicopter

Radial engines are regarded as being air-cooled almost by definition—so that it is interesting that one of the most successful of the early radial engines was the Salmson 9Z series of nine-cylinder water-cooled radial engines that were produced in large numbers during the First World War. Georges Canton and Pierre Unné patented the original engine design in 1909, offering it to the Salmson company—and the engine was often known as the Canton-Unné.

42 cylinder USSR radial engine for warships

Little development of the radial engine was undertaken in Germany during World War I, where most aircraft used water-cooled inline 6-cylinder engines. Two radial engines were made there before the war but they were not proceeded with.[3]

From 1909 to 1919 the radial engine was overshadowed by its close relative, the rotary engine—which differed from the so-called "stationary" radial in that the crankcase and cylinders revolved with the propeller. Mechanically it was identical in concept to the later radial however the prop was bolted to the engine, and the crankshaft to the airframe. The primary reason for this was to ensure cooling of the cylinders, a notorious problem with all of the early radials.

In World War I, many French and other Allied aircraft flew with Gnome, Le Rhône, Clerget and Bentley rotary engines, the ultimate examples of which reached 240 hp (**unknown operator: u'strong'** kW). The German Oberursel firm (who had originated the Gnom design) made licensed copies of the Gnome and Le Rhône powerplants while Siemens-Halske built a number of their own designs including the Siemens-Halske Sh.III eleven-cylinder rotary engine.

By the end of the war the rotary engine had reached the limits of the design - particularly in regard to the amount of fuel and air that could be drawn into the cylinders during the intake stroke due to the rotary motion, while advances in both metallurgy and cylinder cooling finally allowed stationary radial engines to supersede rotary engines. In the early 1920s Le Rhône converted a number of their rotary engines into stationary radial engines although most of the other early radial engines were new designs.

By 1918, the potential advantages of air-cooled radials over the water-cooled inline engine and air-cooled rotary engine that had powered World War I aircraft were well appreciated but remained unrealized. While British designers had produced the ABC Dragonfly radial in 1917, they were unable to resolve its cooling problems, and it was not until the 1920s that the Bristol Aeroplane Company and Armstrong Siddeley produced reliable British radials such as the Bristol Jupiter and the Armstrong Siddeley Jaguar.

In the US, NACA noted in 1920 that air-cooled radials could offer an increase in the power-to-weight ratio and reliability, and by 1921 the US Navy had announced it would only order aircraft fitted with air-cooled radials while other naval air arms followed suit. Charles Lawrance's J-1 engine was developed in 1922 with Navy funding, and using aluminium cylinders with steel liners ran for an unprecedented 300 hours, at a time when 50 hours endurance was normal. At the urging of the Army and Navy the Wright Aeronautical Corporation bought Lawrance's company, and subsequent engines were built under the Wright name. The radial engines gave confidence to Navy pilots performing long-range overwater flights.[5]

Wright's 225 hp (**unknown operator: u'strong'** kW) J-5 Whirlwind radial engine of 1925 was widely acknowledged as "the first truly reliable aircraft engine".[6] Wright employed Giuseppe Mario Bellanca to design an aircraft to showcase it, and the result was the Wright-Bellanca 1, or WB-1, which was first flown in the latter part of that year. The J-5 was used on many advanced aircraft of the day, including Charles Lindbergh's Spirit of St. Louis with which he made the first solo trans-Atlantic flight.

In 1925, the American rival firm to Wright's radial engine production efforts, Pratt & Whitney, was founded. The P & W firm's initial offering, the Pratt & Whitney R-1340 Wasp, test run later that year, began the evolution of the many models of Pratt & Whitney radial engines that were to appear during the second quarter of the 20th century, among them the 14-cylinder, twin-row Pratt & Whitney R-1830 Twin Wasp, the most-produced aviation engine of any single design, with a total production quantity of nearly 175,000 engines.

In the United Kingdom the Bristol Aeroplane Company was concentrating on developing radials such as the Jupiter, Mercury and sleeve valve Hercules radials. France, Germany, Russia and Japan largely built licenced or locally improved versions of the Armstrong Siddeley, Bristol, Wright, or Pratt & Whitney radials.

Radial versus inline debate

Liquid-cooled engines often weigh more and their cooling systems are both more complex and are generally more vulnerable to battle damage. Minor shrapnel damage easily results in a loss of coolant and consequent engine seizure, while an air-cooled radial would be unaffected.[7] Additionally, radials offer higher mechanical efficiency than inline engines, as they have shorter and stiffer crankshafts, a single bank radial needing only two crankshaft bearings as opposed to the seven required for a six-cylinder inline engine of similar stiffness.[8] The shorter crankshaft also produces less vibration and hence higher reliability through reduced wear.

1935 Monaco-Trossi, a rare example of automobile use.

While a single bank radial permits all cylinders to be cooled equally, the same is not true for multi-row engines where the rear cylinders are affected by the heat coming off the front row, and air flow being masked.[9] Additionally, having the cylinders in the airflow increases drag considerably, adding turbulence that destroys the laminar airflow over the fuselage and adjacent wings. The answer to both these problems was the addition of specially designs cowlings with baffles to force the air over the cylinders. The first effective drag reducing cowling that didn't impair engine cooling was the British Townend ring or "drag ring" which formed a narrow ring around the engine covering the cylinder heads, not only reducing drag, but adding a small amount of thrust. NACA then studied the problem further and developed the NACA cowling which further reduced drag, increased thrust and improved cooling. Nearly all aircraft radial engine installations since have used NACA type cowlings. Several aircraft with liquid-cooled engines in service until the end of the Second World War, such as the Spitfire and the P-51 took advantage of the Meredith Effect to generate thrust, which worked in a similar manner to the NACA cowling.[10]

The radial engine typically has a larger frontal area, and is less amenable to streamlining and drag reduction than an inline engine. Pilot visibility is often sacrificed due to the greater width of the engine, and the designer is more limited in engine placement as greater care must be taken to ensure adequate cooling air, either in a buried engine installation or in a pusher configuration.

While inline liquid-cooled engines continued to be common until the end of World War II, radial engines saw widespread service in the successful Mitsubishi Zero and Focke-Wulf Fw 190, while the late-war Hawker Sea Fury and Grumman Bearcat, two of the fastest production piston-engined aircraft ever built, used radial engines. Until the development of the jet engine, large aircraft commonly used radial engines. Factors influencing the choice of radial over inline were reliability, simplicity in maintenance, and the ability to package a radial as a power egg, readily removable from the aircraft with the disconnection of only a few lines. Additionally, the large frontal area of these aircraft meant the radial engine's own frontal profile was a less significant factor when it came to drag.

Multi-row radials

Originally radial engines had one row of cylinders, but as engine sizes increased it became necessary to add extra rows. The first known radial-configuration engine to ever use a twin-row design was the 160 hp Gnôme "Double Lambda" rotary engine of 1912, designed as a 14-cylinder twin-row version of the firm's 80 hp Lambda single-row seven-cylinder rotary, with only the German Oberursel U.III clone of the Double Lambda reproducing the Gnome Double Lambda's twin-row design before the end of World War I. Most stationary radial engines did not exceed two rows, but the largest displacement radial engine ever built in quantity, the Pratt & Whitney R-4360 Wasp Major, with cylinders in *corncob* configuration, was a 28-cylinder 4-row

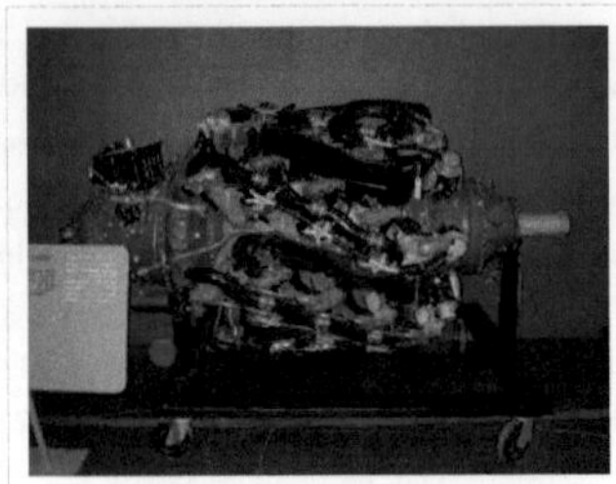

The Wasp Major, a four-row radial.

radial engine used in many large aircraft designs in the post-World War II period. The Lycoming R-7755 was the largest piston-driven aircraft engine ever produced; with 36 cylinders totaling about 7,750 in^3 (127 L) of displacement and a power output of 5000 horsepower (**unknown operator: u'strong'** kW). It was originally intended to be used in the "European bomber" that eventually emerged as the Convair B-36. Only two examples were built before the project was terminated in 1946. The USSR also built a limited number of 'Zvezda' engines with up to 56 cylinders, which were even larger in displacement than the Lycoming R-7755. The 112-cylinder diesel boat engines featuring 16 rows with 7 banks of cylinders, bore of 160 mm (6.3 in), stroke of 170 mm (6.7 in), and total displacement of 383 liters (23,931 in^3). The engine produced 10000 hp (**unknown operator: u'strong'** kW) at 2,000 rpm. They were used on fast attack craft, such as Osa class missile boats.

Modern radials

At least five companies build radials today. Vedeneyev engines produces the M-14P model, 360 hp (**unknown operator: u'strong'** kW) (up to 450 hp (**unknown operator: u'strong'** kW)) radial used on Yakovlev, Sukhoi Su-26, and Su-29 aerobatic aircraft. The M-14P has also found great favor among builders of experimental aircraft, such as the Culp's Special, and Culp's Sopwith Pup [11], Pitts S12 "Monster" and the Murphy "Moose". 110 hp (**unknown operator: u'strong'** kW) 7-cylinder and 150 hp (**unknown operator: u'strong'** kW) 9-cylinder engines are available from Australia's Rotec Engineering. HCI Aviation [12] offers the R180 5-cylinder (75 hp (**unknown operator: u'strong'** kW)) and R220 7-cylinder (110 hp (**unknown operator: u'strong'** kW)), available "ready to fly" and as a build-it-yourself kit. Verner Motor, from the Czech Republic, now builds several radial engines. Models range in power from 71 hp (**unknown operator: u'strong'** kW) to 172 hp (**unknown operator: u'strong'** kW). [13] Miniature radial engines for model airplane use are also available from Seidel in Germany, OS and Saito Seisakusho of Japan, and Technopower in the USA. The Saito firm is known for making three different sizes of 3-cylinder radials, as well as a 5-cylinder example, as the Saito firm is a specialist in making a large line of miniature four-stroke engines for model use in both methanol-burning glow plug and gasoline-fueled spark plug ignition engine formats.

Diesel radials

While most radial engines have been produced for gasoline fuels, there have been instances of diesel fueled radial engines. Two major advantages favour diesel engines - reduced fuel consumption and reduced risk of fires - however they have all the disadvantages to which diesel engines have been prone.

Packard designed and built a diesel radial aircraft engine, the DR-980, in 1928. It was a 9-cylinder radial engine displacing 980 cubic inches and rated at 225 horsepower (**unknown operator: u'strong' kW**). On 28 May 1931, a Bellanca CH-300 fitted with a DR-980, piloted by Walter Edwin Lees and Frederick Brossy, set a record for staying aloft for 84 hours and 32 minutes without being refueled.[14] This record was not broken until 55 years later by the Rutan Voyager.[15]

Packard DR-980 diesel radial aircraft engine.

The experimental Bristol Phoenix of 1928–1932 was successfully flight tested in a Westland Wapiti and set altitude records in 1934 that lasted until World War II.

In 1932 the French company Clerget developed the 14D, a 14-cylinder two-stroke diesel radial engine. After a series of improvements, in 1938 the 14F2 model produced 520 hp (**unknown operator: u'strong' kW**) at 1910 rpm cruise power, with a power-to-weight ratio near that of contemporary gasoline engines and a specific fuel consumption of roughly 80% that for an equivalent gasoline engine. During WWII the research continued, but no engines were

A Nordberg Manufacturing Company two-stroke diesel radial engine for power generation and pump drive purposes.

mass-produced because of the Nazi occupation. By 1943 the engine had grown to produce over 1000 hp (**unknown operator: u'strong' kW**) with a turbocharger. After the war, the Clerget company was integrated in the SNECMA company and had plans for a 32-cylinder diesel engine of 4000 hp (**unknown operator: u'strong' kW**), but in 1947 the company abandoned piston engine development in favor of work on the emerging turbine engines.

The Nordberg Manufacturing Company of the US developed and produced a series of large two-stroke radial diesel engines from the late 1940s for electrical production, primarily at aluminium smelters and for pumping water. They differed from most radials in using having an even number of cylinders in a single bank (or row), thanks to an unusual double master connecting rod that allowed the engine to be timed so the cylinders fired in consecutive order. Variants were built that could be run on either diesel oil or gasoline or mixtures of both. A number of powerhouse installations utilising large numbers of these engines were made in the US.[16]

The Zoche[17] in Germany have produced a prototype range of radial air-cooled two-stroke diesel aircraft engines, comprising a V-twin, a single-row cross-4 and a double-row cross-8.[18] A Zoche engine has run successfully in wind tunnel tests,[19] but Zoche seem barely closer to production than they were a decade ago. Experimental engine manufacturers often experience difficulties in proceeding beyond the prototype stage due to high development and certification costs particularly with markets dominated by relatively cheap WWII era engines.

Compressed air radials

A number of radial motors operating on compressed air have been designed, mostly for use in model airplanes, or simply as a building project. Radial compressed air engines include Liney Machine Halo 5 [20], Mike Smyth SF376 [21], Taiyo UTAM4 [22], ... They have also been used in gas compressors.[23]

Use in tanks

In the years leading up to WWII, as the need for armored vehicles was realized, designers were faced with the problem of how to power the vehicles, and turned to using aircraft engines, among them radial types. The radial aircraft engines provided greater power-to-weight ratios and were more reliable than conventional inline vehicle engines available at the time. This reliance had a downside though: if the engines were mounted vertically as in the M3 Lee and M4 Sherman, their comparatively large diameter gave the tank a higher silhouette than designs using inline engines.

The Continental R-670, a 7-cylinder radial aero engine which first flew in 1931, became a widely used tank powerplant, being installed in the M1 Combat Car, M2 Light Tank, M3 Stuart, M3 Lee, LVT-2 Water Buffalo.

The Guiberson T-1020, a 9-cylinder radial diesel aero engine, was used in the M1A1E1, M2, and M3, while the Continental R975 saw service in the M4 Sherman, M7 Priest, M18 Hellcat tank destroyer, and the M44 self-propelled howitzer.

Model radial engines

A number of multi-cylinder 4-stroke model engines have been commercially available in a radial configuration, beginning with the Japanese O.S. Max firm's FR5-300 five-cylinder, 3.0 cu.in. (50 cm3) displacement "Sirius" radial in 1986. The American 'Technopower' firm had made smaller-displacement five- and seven-cylinder model radial engines as early as 1976, but the OS firm's engine was the first mass-produced radial engine design in aeromodeling history. The rival Saito Seisakusho firm in Japan has since produced a similarly sized five-cylinder radial four-stroke model engine of their own as a direct rival to the OS design, with Saito also creating a trio of three-cylinder radial engines ranging from 0.90 cu.in. (15 cm3) to 4.50 cu.in. (75 cm3) in displacement. The German Seidel firm has made both seven- and nine-cylinder "large" (starting at 70 cm3 displacement) radio control model radial engines, mostly for glow plug ignition, with an experimental fourteen-cylinder twin-row radial being tried out.

See also

- Kinner B-5
- List of aircraft engines
- Rotary engine
- Technopower
- Vedeneyev
- Warner Scarab
- Zvezda M503, a 42-cylinder Soviet missile boat diesel radial engine.

References

[1] "Firing order: Definition from" (http://www.answers.com/topic/firing-order). Answers.com. 2009-02-04. . Retrieved 2011-12-06.

[2] http://commons.wikimedia.org/wiki/File:Radial_engine_timing.gif

[3] Vivian, E. Charles (1920). *A History of Aeronautics* (http://www.daytonhistorybooks.citymax.com/page/page/3259323.htm). Dayton History Books Online. .

[4] Day, Lance; Ian McNeil (1996). *Biographical Dictionary of the History of Technology*. Taylor & Francis. p. 239. ISBN 0-415-06042-7.

[5] Bilstein, Roger E. (2008). *Flight Patterns: Trends of Aeronautical Development in the United States, 1918–1929*. University of Georgia Press. p. 26. ISBN 0-8203-3214-3.

[6] Herrmann, Dorothy (1993). *Anne Morrow Lindbergh: A Gift for Life*. Ticknor & Fields. p. 28. ISBN 0-395-56114-0.

[7] Thurston, David B. (2000). *The World's Most Significant and Magnificent Aircraft: Evolution of the Modern Airplane* (http://books.google.com/?id=7HTPRym0iYIC&pg=PA155). SAE. p. 155. ISBN 0-7680-0537-X. .

[8] Some six-cylinder inline engines used as few as 3 bearing but at the cost of heavier crankshafts, or crankshaft whipping.

[9] Fedden, A.H.R. (28 February 1929). "Air-cooled Engines in Service" (http://www.flightglobal.com/pdfarchive/view/1929/1929 - 0433.html). *Flight* **XXI** (9): 169–173. .

[10] Price 1977, p. 24.

[11] http://culpsspecialties.com

[12] http://www.hciaviation.com/

[13] http://www.vernermotor.com/

[14] Aircraft Engine Historical Society - Diesels (http://www.enginehistory.org/Diesels/CH1.pdf) Retrieved: 30 January 2009

[15] Aviation Chronology (http://www.aerofiles.com/chrono.html) Retrieved: 7 February 2009

[16] "Nordberg Diesel Engines" (http://www.oldengine.org/members/diesel/Nordberg/Nordmenu.htm). OldEngine (http://www.oldengine.org). . Retrieved 2006-11-20.

[17] "zoche aero-diesels homepage" (http://www.zoche.de/). Zoche.de. . Retrieved 2011-12-06.

[18] "Prospekt 2007.indd" (http://www.zoche.de/zoche_brochure.pdf) (PDF). . Retrieved 2011-12-06.

[19] "zoche aero-diesels testbench video" (http://www.zoche.de/Zoche_video.html). Zoche.de. . Retrieved 2011-12-06.

[20] http://www.lineymachine.com/lineyhalokit-p-2699.html?osCsid=df1711b4a444c2df252d0dd2684d6802

[21] http://home.ctlnet.com/~robotguy67/classic_cars/air_engines/V-Twin/air_engines.htm

[22] http://www.automation-dfw.com/pdf_pneu/taiyo-01utam4airmotors.pdf

[23] "Bock radial piston compressor" (http://www.bock.de/en/Product_overview.html?ArticleSizesGroupID=169). Bock.de. 2009-10-19. . Retrieved 2011-12-06.

External links

- Rotec radial engines (http://www.rotecradialengines.com/)
- Vedeneyev engines (http://www.russianaeros.com/vedenyevproduct.htm)
- Murphy Moose M-14P info (http://www.murphyair.com/Product_Info/Super/M-14.htm)
- Radial Engine Motorcycles (http://www.jrlcycles.com/page/page/4187437.htm)
- Radial Engine Automobile (http://www.deutsche-werke.de/goggo2.htm) (German)

Article Sources and Contributors

Renard_R.35 *Source*: http://en.wikipedia.org/w/index.php?title=Renard_R.35 *Contributors*: Ahunt, Chris the speller, MilborneOne, Nigel Ish

Airliner *Source*: http://en.wikipedia.org/w/index.php?title=Airliner *Contributors*: A380 Fan, Acroterion, Ahoerstemeier, Ahunt, Airplaneman, Al Lemos, Alexius08, AndersL, Andrew Duffell, AndrewHowse, Andros 1337, AniRaptor2001, Antemeridian, AntonioMartin, Arpingstone, Art LaPella, BHenry1969, BRG, Banstaman, Bapho, Beland, BilCat, Blimpguy, Bobblewik, Borgx, Bryan Derksen, Bzuk, CSWarren, Camerafiend, Camerong, Canterbury Tail, Ccacsmss, Chris the speller, Clawson, CommonsDelinker, CopperMurdoch, Correctemundo, Cricketseven, Cromulent, CzarB, Dan100, David.Monniaux, DeweyQ, Dewritech, DexDor, DocWatson42, Dpm64, Dschwen, Dskluz, Dssis1, Dtom, EditorASC, Edward, Egil, El C, Enbob89, Ennerk, Eregli bob, Ericg, Erielhonan, Espoo, Excirial, F.bendik, Falconus, Fanatix, Fastardul, Ff1959, FlyerBoy, Fnlayson, Funandtrvl, GeneralCheese, Gogo Dodo, GraemeLeggett, Greyhood, Grhornik, HIST406-10rlavoie, Hans Dunkelberg, Harizhazman, Hatcat, Henning Makholm, Hoeksas, IRelayer, Icairns, Inetpup, Ixfd64, J.delanoy, Jam01, Jan olieslagers, JetBlast, Jfsa380, Jguk, John, John Barleycorn's Revenge, JonathanDP81, Koavf, Kozuch, Ksyrie, LOL, Lightmouse, Linuxbeak, M-le-mot-dit, MIT Trekkie, Magioladitis, Manassehkatz, Marsian, Mattfalcus, Maxzden, Megane, Mike1942f, MilborneOne, MinorEdit, Mion, MisfitToys, Mohammed Hamadiyah, Mono, Murray Langton, Mythrilfan, Nakon, Olivier, Ospalh, Outriggr, Ouyuecheng, OverlordQ, Patrick, Pearle, Peter Horn, Pj747, Planetary Chaos, PolarYukon, Ppntori, RTC, Random Tree, Rcav8rs, Reedy, Rettetast, Rich Farmbrough, Rjwilmsi, Rlandmann, Rmhermen, RobertG, RonH, RottweilerCS, Russavia, SCDBob, SarekOfVulcan, Sekicho, SilkTork, Simone, Sjakkalle, Sjorford, Slii, Smalljim, Soarhead77, Sp33dyphil, Spartan7W, Spellmaster, SrAtoz, SteamSpeed, Stephenchou0722, StoneProphet, Sukee3, Sum0, Tagremover, Tamariki, Tangotango, Tannin, Tatrgel, Tdadamemd, Tgeairn, TheFeds, Thegraham, Thumperward, Tomccoll, Turidoth, Urban Datura, Vance&lance, Vaughan Pratt, Vdlsnes, Vgy7ujm, Voyevoda, Watermaren, Wernher, WhisperToMe, Whkoh, Who, Wimt, WoodenTaco, Wrightbus, XJamRastafire, Yourworldelivered, ZS, Zalgo, Zeeshan-NJITWILL, Zero76, 386 anonymous edits

Monoplane *Source*: http://en.wikipedia.org/w/index.php?title=Monoplane *Contributors*: 7&6=thirteen, Abmac, Ahunt, Airwolf, Alma Pater, Andrew Iverson, Anittas, Arpingstone, Arrivisto, Autodidactyl, Ballon of pi, Benstown, Berserkerus, BilCat, Breez, Chochopk, Chris the speller, CommonsDelinker, Coosbane, DanMS, Darkwind, DigitalGhost, DrKiernan, Dtom, Eloy, Empowered, Enlil Ninlil, FiggyBee, Glacialfox, GliderMaven, Hohum, J.delanoy, Japanese Searobin, Joergen, Koplimek, Kubanczyk, La goutte de pluie, Lkinkade, Ludovic89, MER-C, Mactaveous, Nibios, P199, Per Honor et Gloria, Pibwl, Plasticbadge, Rcingham, Rlandmann, RottweilerCS, Salsa Shark, Sharkface217, Soundofmusicals, Styrofoam1994, Subversive.sound, Sumsum2010, The High Fin Sperm Whale, The PIPE, Tigga en, Tirkfl, TraceyR, XJamRastafire, Yekrats, 44 anonymous edits

Sabena *Source*: http://en.wikipedia.org/w/index.php?title=Sabena *Contributors*: 7472u3, AMCKen, AWeenieMan, AdAstra reloaded, Adhominem, Ahoerstemeier, Airtimes, AlbertR, Anas1712, Andypasto, Apoivre, Ardfern, Arent1, ArglebargleIV, Arpingstone, Astrotrain, Attilios, B-west, BD2412, Badger151, Bethling, Bobblewik, Bobo192, BokicaK, Butakun, CLW, Caiaffa, Calvin Wilton, CambridgeBayWeather, Chesipiero, Chilledsunshine, Choster, Chowbok, Chris the speller, ChrisCork, Cj1340, Clipper471, Closeapple, Cobatfor, Colonies Chris, CommonsDelinker, Dantadd, DavidWBrooks, DennisJOBrien@yahoo.com, Dlom, Donald Albury, Drdanny, Dricherby, Duncan7670, Dupont och Dupond, Eassa, Eggman64, Error, Evaluist, Fandoo, Fdewaele, Ffransoo, Flymeoutofhere, GSMartin, Gaius Cornelius, Gamsbart, Gittinsj, Greyengine5, Gryffindor, Gsandi, HMSSolent, Hadal, Hailey C. Shannon, Henri Musielak, Huaiwei, Hydraton31, Ibrahim1726, JHunterJ, JamesMLane, Jamesluckard, Jan olieslagers, Jan.volker, Jangli, Jason M, Jevansen, Jim10701, JohnOwens, Jovianeye, Justinbb, Kabila, Kate, Kenny Moens, Kjd, Klaus Vomhof, Klemen Kocjancic, Klow, Kusma, Lkinkade, Lstanley1979, LukaP, MKY661, Malikbek, Malo, Markonen, Mdnavman, Mikeblas, Mikenucklesii, MilborneOne, Milliped, Mion, Mjroots, Morven, Mtaylor848, Neep, Nick Mks, Nicolaiplum, Noelmg, Nono64, Onovase, Pagrashtak, Peterleerocks, Pierre.becquart, Pitlane02, Pol098, Pvosta, Rbanzai, Rettetast, Rich Farmbrough, Richard Arthur Norton (1958-), Richsage, Ringwayobserver, Rjwilmsi, Russavia, RuthAS, SABENA, SNIyer12, Sardanaphalus, Searcher 1990, Sesel, Shultzc, Sonett72, Sousclef, Sp33dyphil, Ssolbergj, Stevenmitchell, TJRC, TahitiB, Tamfang, Tedernst, Teles, Template namespace initialisation script, Thenoflyzone, Tijuana Brass, Tlotoxl, Tobibln, Troutsneeze, Ttwaring, Urhixidur, VanBeem, Vlad, Vremya, Wangi, WhisperToMe, Wikifunkstar, Wim Devolder, Woohookitty, Worcsdave, XLR8TION, 234 anonymous edits

Belgian_Congo *Source*: http://en.wikipedia.org/w/index.php?title=Belgian_Congo *Contributors*: 79spirit, APB-CMX, Adam Scott 89, Adelbrecht, Ahoerstemeier, Ali@gwc.org.uk, Amr2323, Andres.geurts, Andrwsc, Antandrus, Aranel, B1mbo, BadLeprechaun, Barryob, Bastin, Belovedfreak, Beneaththelandslide, Bfigura's puppy, BigHairRef, Bigblue800, Bigblue801, Brettreasure, Brhaspati, Btsz, Buistr, Buster7, Canterbury Tail, Catherine Huebscher, Charles Matthews, Chasnor15, Colonies Chris, CommonsDelinker, Conte di Cavour, CopperSquare, Cpastern, Cs-wolves, DWC LR, Danensis, DanielCD, Den fjättrade ankan, Design, Dionysos1, Discospinster, Dlohcierekim, DocWatson42, Domino theory, DougsTech, Dvyost, Edcolins, Efghij, Ehrenkater, Eiland, Eltomzo, Ezeu, Fainites, Fastfission, Fieldday-sunday, Finell, Flix11, Frangibility, Frietjes, Friginator, Frumentarius, Ghirlandajo, Gimboid13, Golbez, Good Olfactory, Gozar, Greenshed, Ground Zero, Grutness, Hawt-babi3, Hektor, Hmains, Hooiwind, Ian Cheese, Iridescent, J.delanoy, JLogan, JSquish, Japanese Searobin, Jim.henderson, JoSePh, John, John K, John O'C, JoshuaWalker, Jspence58, Juliancolton, Kalathalan, Karan Kamath, Karl Stas, Kelvinc, Khazar, Kintetsubuffalo, Kreetslaak, Kwamikagami, La8470a, Laurinavicius, Leszek Jańczuk, Lights, Lihaas, Listmeister, Look2See1, LorenzoB, MZMcBride, Magister Mathematicae, Mahanga, Mangostar, Manxruler, Maralia, Marktreut, Martin Wisse, MatthewVanitas, Mav, MazabukaBloke, Mboverload, McGoo, Milesli, Mimihitam, Mkpumphrey, Moink, MptsWiki, Muhends, Nach0king, Nikolainz, Nk, Nnemo, No Guru, Ohconfucius, Oldknock, PL290, Parker j kukral, Paul August, Pearle, Peripatetic, Peter Karlsen, Pharos, PhilKnight, Philip Trueman, PhilipC, PhnomPencil, Pibwl, Piet Clement, Polylerus, Pprevos, Pseudoanonymous, Psycho Kirby, Publunch, Puffin, Pvosta, Ragib, Raymondwinn, Rex Germanus, Rich Farmbrough, Richhoncho, Rjwilmsi, RobertG, Rocastelo, Sadads, Sarefo, Sce03066, ScopyCat, Smith2006, Snoyes, Some jerk on the Internet, Str1977, Sugarbat, Suidafrikaan, Superslum, Supspirit, Tabletop, Tad Lincoln, Terry J. Carter, Tgeairn, The Epopt, The Madras, The Moose, The wub, Themalau, Thom2002, Thomas Ryckmans, Timbobutcher, Tommy2010, Triddle, Triwbe, Ugen64, User27091, Van helsing, Vberger, Vbrems, Velho, Vipinhari, Vito Genovese, Vzbs34, Waide Piki, Wester, Whiskey Pete, WhisperToMe, Wiki alf, Wildaker, Woohookitty, Wozzow, Xed, Xod, Yulia Romero, Zscout370, Україна2000, Уральский Кот, 343 anonymous edits

Gnome-Rhône_Mistral_Major *Source*: http://en.wikipedia.org/w/index.php?title=Gnome-Rh%C3%B4ne_Mistral_Major *Contributors*: AMCKen, Ain92, Andrewa, BilCat, Cobatfor, Denniss, Emt147, Gampe, GraemeLeggett, Idsnowdog, Maury Markowitz, Nigel Ish, Nimbus227, Petebutt, PpPachy, Rgvis, Rlandmann, TSRL, Trekphiler, Wxtype, 36 anonymous edits

Gnome-Rhône_9K *Source*: http://en.wikipedia.org/w/index.php?title=Gnome-Rh%C3%B4ne_9K *Contributors*: AMCKen, Ain92, Carabinieri, Chris the speller, Idsnowdog, Nimbus227, PpPachy, Rlandmann, Sardanaphalus, TSRL, Threecharlie, Trekphiler, Ygolovk, 6 anonymous edits

Radial_engine *Source*: http://en.wikipedia.org/w/index.php?title=Radial_engine *Contributors*: AGToth, AMCKen, Aeons, Alstrupjohn, Andrew Nutter, Andrewa, Ankon Ray, Arjayay, Arnero, Arrivisto, Atlant, Axeman89, Bagheera, Balloonguy, Bernzoil, Bigjimr, BilCat, Blandoon, Brianhe, Bryan Derksen, Chris the speller, Colin Douglas Howell, Daleh, Davygrvy, Dhollm, Dori, Dpwkbw, Duk, Ed g2s, Emt147, Encephalon, Enzedrail, Ericd, Ericg, Euchiasmus, F.bendik, Fantastic4boy, FrummerThanThou, GCarty, Geni, Glenn, Gogtjop, Gooberliberation, GraemeLeggett, Gregorio Damian Gomez, Greyengine5, Hellbus, Hooperbloob, Howcheng, Humphrey20020, Huw Powell, Iancarine, Jmc41, Joevsimp, Joffeloff, Jóna Þórunn, Kaisershatner, Kendrick7, Kieff, Kingernie, Knotnic, Kubanczyk, Lahiru k, Leandrod, Lexington50, Liftarn, Lightmouse, LilHelpa, Lizardo tx, Lockesdonkey, Longhair, Man91816, Mark.murphy, MartinezMD, Maury Markowitz, Mccraw274, Mhrogers, Michael Daly, Michael Frind, MisfitToys, Moletrouser, Morven, Moshe Constantine Hassan Al-Silverburg, Mulad, Neodarkshadow, NiD.29, Nimbus227, Nuance 4, Ospalh, Oxhop, Paul Richter, Pavithran, Petri Krohn, Pibwl, Pol098, Polpo, Psb777, Richard Arthur Norton (1958-), Rock4arolla, Rogerd, RottweilerCS, Sabedon, Salmanazar, Sardanaphalus, Segv11, Serasuna, Sfoskett, Shreditor, Siqbal, Softeis, Soundofmusicals, Stoianovici, TSRL, Template namespace initialisation script, The PIPE, TimVickers, Trekphiler, Trieste, Typ932, Vykk, Werdan7, Wizzy, Wtmitchell, Xiner, Yonatan, Youngjim, Ævar Arnfjörð Bjarmason, 164 anonymous edits

Image Sources, Licenses and Contributors

File:Congo spoorwegen.gif *Source*: http://en.wikipedia.org/w/index.php?title=File:Congo_spoorwegen.gif *License*: unknown *Contributors*: Albert Sarlet

File:White nurses of the Union Minière du Haut Katanga April 1918.JPG *Source*:
http://en.wikipedia.org/w/index.php?title=File:White_nurses_of_the_Union_Minière_du_Haut_Katanga_April_1918.JPG *License*: unknown *Contributors*: Gourdinne

File:Albert Militair Kamp Leopoldstad.JPG *Source*: http://en.wikipedia.org/w/index.php?title=File:Albert_Militair_Kamp_Leopoldstad.JPG *License*: unknown *Contributors*: Koninklijk Paleis

File:Kisantu kathedraal 2.JPG *Source*: http://en.wikipedia.org/w/index.php?title=File:Kisantu_kathedraal_2.JPG *License*: unknown *Contributors*: User:PClement

File:Lubumbashi Palais de Justice 2.jpg *Source*: http://en.wikipedia.org/w/index.php?title=File:Lubumbashi_Palais_de_Justice_2.jpg *License*: unknown *Contributors*: Piet Clement

File:PatricelumumbaIISG.jpg *Source*: http://en.wikipedia.org/w/index.php?title=File:PatricelumumbaIISG.jpg *License*: unknown *Contributors*: Collection IISG

File:Gnome-Rhone 14 Mistral Major engines 1943.jpg *Source*: http://en.wikipedia.org/w/index.php?title=File:Gnome-Rhone_14_Mistral_Major_engines_1943.jpg *License*: unknown *Contributors*: U.S. Army Signal Corps

Image:Radial engine.gif *Source*: http://en.wikipedia.org/w/index.php?title=File:Radial_engine.gif *License*: unknown *Contributors*: User:Duk

Image:Radial engine WACO QCF2.jpg *Source*: http://en.wikipedia.org/w/index.php?title=File:Radial_engine_WACO_QCF2.jpg *License*: unknown *Contributors*: User:TimVickers

File:TwinWaspConRods.jpg *Source*: http://en.wikipedia.org/w/index.php?title=File:TwinWaspConRods.jpg *License*: unknown *Contributors*: User:TSRL

Image:Rotary Piston Engine 8b03632r.jpg *Source*: http://en.wikipedia.org/w/index.php?title=File:Rotary_Piston_Engine_8b03632r.jpg *License*: unknown *Contributors*: Andy Dingley, Denniss, Marcelloo, Nimbus227, Saperaud, Sfoskett, Stahlkocher, 1 anonymous edits

Image:H19 showing engine.jpg *Source*: http://en.wikipedia.org/w/index.php?title=File:H19_showing_engine.jpg *License*: unknown *Contributors*: Alaniaris, Marcelloo, Rogerd, Stahlkocher, Threecharlie

File:USSR ship radial engine.JPG *Source*: http://en.wikipedia.org/w/index.php?title=File:USSR_ship_radial_engine.JPG *License*: unknown *Contributors*: User:Neodarkshadow

Image:Monaco-Trossi1935.jpg *Source*: http://en.wikipedia.org/w/index.php?title=File:Monaco-Trossi1935.jpg *License*: unknown *Contributors*: Andy Dingley, Herranderssvensson, Kobac, Liftarn, Sporti

Image:Pratt & Whitney R-4360 Wasp Major 1.jpg *Source*: http://en.wikipedia.org/w/index.php?title=File:Pratt_&_Whitney_R-4360_Wasp_Major_1.jpg *License*: unknown *Contributors*: User:Highflier

File:Packard DR-980 USAF.jpg *Source*: http://en.wikipedia.org/w/index.php?title=File:Packard_DR-980_USAF.jpg *License*: unknown *Contributors*: Kogo, Nimbus227, Stahlkocher

Image:IMG 0648.JPG *Source*: http://en.wikipedia.org/w/index.php?title=File:IMG_0648.JPG *License*: unknown *Contributors*: User:DanielHolth

Printed by Books on Demand GmbH, Norderstedt / Germany